The Television PA's Handbook

The Television PA's Handbook

Second edition

Avril Rowlands

Focal Press
An imprint of Butterworth-Heinemann Ltd
Linacre House, Jordan Hill, Oxford OX2 8DP

A member of the Reed Elsevier Group

OXFORD LONDON BOSTON
MUNICH NEW DELHI SINGAPORE SYDNEY
TOKYO TORONTO WELLINGTON

First published as *The Production Assistant in TV and Video* 1987
Second edition 1993

British Library Cataloguing in Publication Data
Rowlands, Avril
Television PA's Handbook. – 2Rev. ed
I. Title
791.43

ISBN 0 240 51353 3

Library of Congress Cataloguing in Publication Data
Rowlands, Avril.
The television PA's handbook/Avril Rowlands. – 2nd ed.
p. cm.
Rev. ed. of: The production assistant in TV and video. 1987.
Includes index.
ISBN 0 240 51353 3
1. Television – Production and direction. 2. Video recordings – Production and direction. 3. Television broadcasting – Great Britain. I. Rowlands, Avril. Production assistant in TV and video. II. Title.
PN1992.75.R6 1993 93–3718
791.45'0232–dc20 CIP

Composition by Genesis Typesetting, Rochester, Kent
Printed in Great Britain by Clays Ltd, St Ives plc

Contents

Part Two: Setting Up

Part Four: Post-Production

Part Five: The Wide World of Television

Preface

Since publication of my book, *The Production Assistant in TV and Video* in 1987, the broadcasting industry in the UK has undergone two revolutions, one technological and the other political.

The industry has witnessed – and is still witnessing – technological change on a massive scale, especially in the areas of videotape formats, post production editing, stereo sound, and computerized graphics and video effects. By the time this revised book reaches the shops it will already be outdated in some respects. I can hazard a guess that there will be at least one or maybe two new digital videotape formats on the market.

Allied to this technological revolution has come a major reshaping of broadcasting in this country. The monopoly of the BBC and the independent television companies within the ITV framework is swiftly becoming a thing of the past. For over thirty years the system has been that with the exception of bought-in programmes from overseas, the television companies in this country were producers and broadcasters. Programmes were made, in the main, by their own permanent staff using their own production facilities. All this has changed, in some instances quite dramatically, during the past five years.

The implications of these changes have led to an increasing casualization on the part of those working in the industry. When I originally wrote this book, the freelance PA was something of a rarity – enough of one to devote a separate chapter to the strange phenomenon! I would now estimate that the freelance PA is more the norm rather than the exception and that this trend is set to continue, even within the confines of the BBC. No longer can anyone working in the industry expect a job for life, working for one company.

So things are not easy out there in today's world of television. No longer are jobs guaranteed, but the very role of the PA, which used to be quite clearly defined, may alter, depending on the size and set-up of the company for which she currently works. In some companies the PA has disappeared altogether, while others, in the interests of cost cutting, have been practising some strange economies such as no longer taking the PA on location.

Very often today's PA will not have the secure umbrella of a large organization around her, with people of experience to whom she can turn if any difficulty arises. Neither can she hide behind years of custom and practice, with established departments set up for the booking of artists, the clearing of copyright etc. Today's PA might be required to set things up for herself, accept that there are no longer any ground rules and work out the goalposts of her job on her own. That does not mean, however, that in this brave new world of broadcasting anything goes, although inherently there is that danger. If that were to happen, over forty years of experience of programme-making would be lost and broadcasting standards would rapidly deteriorate. It is therefore more, not less, essential for the PA to be knowledgeable about the job. It is more essential for her to be versatile in coping with different types of programme and different types of company. It is more essential for her to be professional in her approach and have the ability of offering the specialized skills of both studio and location work.

While my first book was written, by and large, for the PA who knew what the job was about, working in an industry within a set and fairly rigid structure, this revised edition is written for those with very different needs:

- For those who come into the industry as production secretaries or general dogsbodys and who are suddenly pitchforked into coping with the specialized skills of the PA with no form of training other than the glib assurance that 'studios are simple'!
- For those who would like to enter the industry but who, possibly after years of study at college or polytechnic following an industry-related course, are unsure about the various options available or about what a PA is or does in television.
- For those who work outside the broadcast industry but in related fields such as corporates, whose work mirrors that of the broadcast PA.
- It is also written for those PAs whose previous jobs with large companies have disappeared and who are now struggling in the uncertain world of the freelance. For those who perhaps specialized in the past in one area of the PA's responsibilities and now need refreshing and updating on other areas.

For all those and for anyone who picks up the book with no other thought that gaining a 'behind-the-scenes' glimpse at how a television programme is made, I hope that this book will be both a help and a guide.

Inkberrow,
Worcestershire

Acknowledgements

I am indebted to many individuals and companies for assistance with research into this book. Among them I would like to thank the Independent Television Companies Association for their generous support; also staff at BBC Pebble Mill, BBC Outside Broadcasts, Coventry Cable, Granada Television, ITN, Limelight Productions, Limehouse Studios, LWT, Shell Video Unit, Sky Channel, Sony Broadcast, Thames Television, TV-AM, TVS and the University of Birmingham Film Unit. I cannot list the many people from whom I received help, but I am most grateful to them all. For this revised edition, I would like to express my thanks to the staff of BBC 'Midlands Today' for allowing me to spend a day with them.

Finally, my love and thanks to my husband, Christopher, who has patiently read this revised edition. I am indebted to him for the new chapters on Videotape editing and The PA's guide to videotape.

Introduction

There you sit, in the darkened control gallery of the studio, a pencil grasped tightly in your slightly damp and decidedly nervous hand, your eyes fixed on the running order as you try to make sense of the row of figures down the page.

On one side of you sits the producer who leans across, wafting unmistakeable signs of stress – and monopolizes your space, your nose and your telephone.

On your other side, the director twitches uncontrollably and picks up one of your four stopwatches (one master watch, one insert and two 'just in case'), thus displacing the neatly regimented row which you had just laid out. You feel a moment's irrational anger, elbow the producer out of the way and offer the director a mint humbug which he accepts with all the desperation of a cigarette smoker in a designated 'no smoking' area.

You return to your figures, but your concentration is gone and you replace your stopwatch – having wrested it from the director – into its former position and re-align it together with your running order, your script, blank sheets of paper, ruler, eraser, coloured pens, pencils and the three other watches, into the rigidly precise pattern which is your attempt to impose security upon an insecure environment.

You raise your eyes and are confronted by a battery of television monitors. The layout of this gallery differs from your last job and you have difficulty working out which monitor is which. The sight is so confusing that you return to your figures.

'How much' hisses the producer, leaning over your shoulder. You edge away, but that brings you into too close a proximity with the director so you move back.

'How much what' you ask, hoping he will lose interest in the question as you have not yet worked out the answer.

Unfortunately he persists. 'Over or under?'

You make a rapid calculation. 'Er . . . two minutes over . . . ' you say, hoping not to be pressed for a more precise reply.

The producer studies your running order and decides to cut down on the time allotted to the first interview. You tell the presenter, thereby incurring the wrath of the director who is working out some complicated digital video effects with the vision mixer and dislikes others talking at the same time. You return to your timings then glance up at the studio clock.

'Five minutes to transmission, five minutes', you say.

The floor manager informs the director that the Important Person who is to be interviewed as the first item has not yet arrived. The floor assistant is sent hot-foot to find out what has happened while the director relieves his feelings by swearing at you, at interviewees who are late and at the world in general.

Reflecting fleetingly upon the Jekyll and Hyde character of directors for whom you have worked, you fail to notice the flashing white light.

'Telephone,' says the studio supervisor sourly. He is a recent non-smoking convert and still suffers the pangs of withdrawal at moments of tension, making him both bad-tempered and irritable.

You answer the telephone and are informed that the Important Person has just arrived at reception in an advanced state of intoxication. You pass on the message and are rewarded by more invective from the director and a homily from the studio supervisor on the evils of drink – he has given up drink as well as cigarettes. The Aston operator shakes her head and carries on typing in the name superimpositions for the programme.

'Are we going to rehearse the opening?' asks an anxious voice, which is disregarded by the director as he is trying out some more complicated electronic effects for possible use later on in the programme.

You, meanwhile, are fully engrossed with marking up fresh pages of script which have just arrived, altering your timings and taking a message from an over-anxious videotape operator who is worried that the insert for Sequence 10 has not arrived. As the insert is not needed until the second half of the programme, you provide the necessary reassurance. Then you hear from presentation control that your programme is to go on the air earlier than planned. Your heart begins to pound and your mouth goes dry as you glance at the studio clock.

'Two minutes to transmission, two minutes.'

'Have to speak up love,' says a disembodied voice.

The director abandons the electronic effects in favour of the more pressing need to rehearse the opening titles. The floor manager reports that the Important Person is being sick in the dressing room and the director abandons the opening titles in order to confer with the producer in view of the interviewee's state of health.

You suddenly notice a switch in front of you on the console desk which you had not noticed before. It has the remains of a label, its title worn away by countless pressing fingers. What is it? Are you supposed to press it, and, if so, when? These thoughts vie in your mind with the conversation taking place between the director and producer and the final calculations you are still making to the programme timing.

The director suddenly suggests starting with Sequence 6. That is vetoed by the producer who suggests Sequence 10. You – sandwiched uncomfortably in the middle – point out that the insert required for Sequence 10 is not yet ready. The director asks why not and you avoid answering by suggesting that you start with Sequence 3. The suggestion is ignored. The director jabs his finger down your running order, the producer digs you in the ribs and, feeling rather like the Dormouse at the Mad Hatter's tea-party, you glance at the studio clock.

'One minute to transmission, one minute.'

What *is* that switch, you wonder frantically.

The director decides to start with Sequence 3 and is just informing the studio when the wan figure of the Important Person is to be seen in the monitor. A joyful floor manager guides him to his seat, gently sits him down and tenderly clips a microphone to his tie.

Butterflies are doing a war dance in your stomach and you fight down an urgent desire to go to the lavatory.

'Stand by VT with opening titles.'

You pick up two of your stopwatches. They feel cold in your clammy hands.

'What are we starting with?' wails the teleprompt operator.

'Thirty seconds to transmission, thirty seconds. Standby studio, standby VT, standby Aston.'

The director runs through the opening sequences.

'What about the red light?' says the sour voice of the studio supervisor. 'The PA died or something?'

You take a deep breath and press the unknown switch. Mercifully, it puts on the red transmission light. You lift your eyes to the studio clock and the transmission monitor which is showing the closing credits of the preceding programme.

'Fifteen seconds.'

There must be an easier way of earning a living, you think, as the adrenalin pumps through your body and your heart beats in slow, heavy thumps.

'There you sit, cool, competent, in control. A complete Production Assistant.'

'10 . . . 9 . . . 8 . . . 7 . . . 6 . . . ROLL VT! . . . 4 . . . 3 . . . 2 . . . 1 . . . ZERO'

You start both stopwatches.

'On opening titles for . . . '

The butterflies have ceased, your hands are steady as you place your master stopwatch on the desk and pick up a red pen. Your voice is calm and clear.

There you sit, cool, competent, in control. A complete Production Assistant.

Part One: General

1 What is a PA?

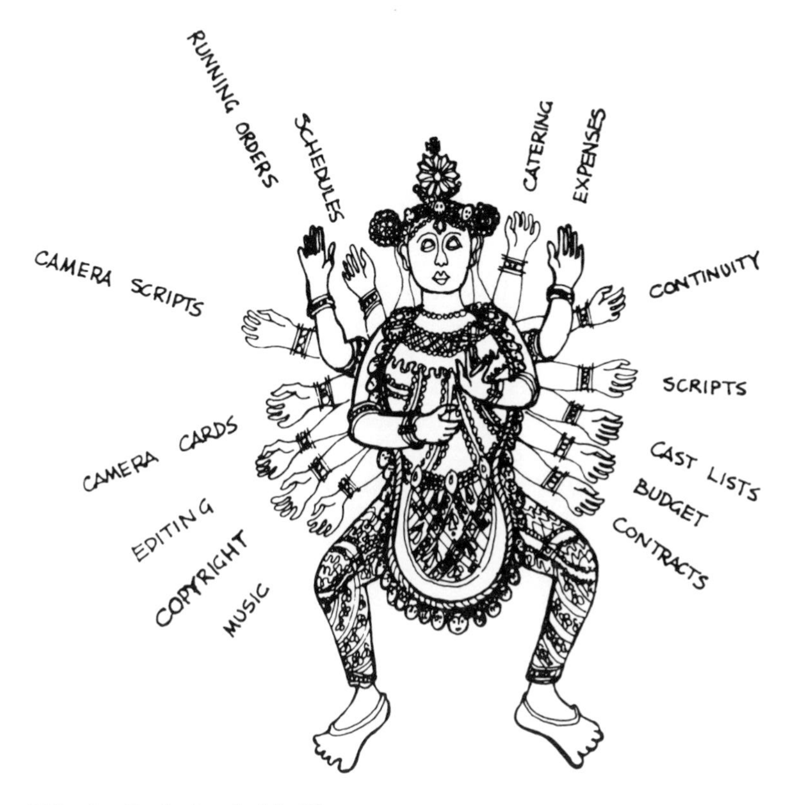

'What is a Production Assistant?'

A Production Assistant is someone who works in the television or the video industry as an essential part of the production team. The PA is, in essence, the director's personal assistant and is usually involved at every stage of production from start to finish. It is a demanding, challenging and immensely satisfying job as it lies at the organizational heart of any production.

In a small company, the PA's job can encompass that of a production manager as well as production secretary and researcher. The PA working in a studio gallery might be required to operate a caption generator, type the script for the teleprompt machine and run in the videotape inserts in addition to carrying out her more traditional role.

The job can comprise some or all of a number of different elements which can very roughly be divided into three separate strands:

Organizational/public relations

The organizational role of the PA means acting as the administrative core of the production – the spider at the centre of the web. The PA is frequently employed at the start of the production, involved in all stages of setting up, and is often the last person remaining, involved in the final stages of post production and clearing up.

This might not be true of all productions. In some companies, the PA might just be engaged for the production itself, exercising her gallery or location skills. Likewise, the larger the production the more people involved. But there is generally a greater or lesser amount of administration and organization required in a PA's job, hand in hand with good secretarial skills. The organizational side of a PA's job will overlap the other strands of her work and requires no specialized training other than that for any good secretary, although she will have to have sufficient technical knowledge in order to make decisions and book the correct facilities.

An ability to liaise and communicate with people on all levels is essential for a PA. She will be required to smooth ruffled feelings and act as a calming influence when egos are damaged and feelings run high.

Location work on film or video

The second strand to a PA's job is that of continuity when involved in out-of-sequence, single-camera shooting, either on film or videotape.

The art of continuity is highly specialized. It is fully covered in my book, *Continuity in Film and Video*, published by Focal Press, 1989.

Studio work

The third strand to the job is the vital role played by the PA in the control room or gallery of a studio when working on live and recorded programmes.

This involves a certain technical knowledge and expertise, fast and accurate time-keeping and the ability to work under intense pressure. The specific tasks of timing, shot calling, rolling in pre-recorded inserts, time-code logging and general communications are skills that are specifically related to this area of the PA's work and, as such, have to be learnt before the PA can cope with a studio gallery unaided. There is also the organizational side of studio work, the camera scripts, running orders, camera cards etc. which need to be understood.

This book focuses especially on this third strand of the PA's job.

Note

Throughout the book I have referred to the Production Assistant as *she*. This is not for any sexist reason but because in the UK the job is more commonly done by a woman. As there is not a suitable pronoun referring equally to *he* and *she*, I hope I will be forgiven for that and for any other apparent stereotypes. They are not intended.

2 A little bit of history

The early days

In these days when the majority of people have video recorders in their own homes, it is important to remember that for many years after the British Broadcasting Corporation started the first public television service in the world in 1936, all programmes were either live or on film and were, of course, in black and white, as were most cinema feature films except for the most lavish.

Live television

Nowadays we tend to take television for granted. The fact that we can see, from the comfort of our armchair, a football match that is being played at that moment on the other side of the world and conveyed to us by satellite is something we all accept. But in the early days it was thought of as something totally magical and its popularity spread rapidly.

To begin with almost everything was transmitted live, not just the programmes we are now accustomed to seeing live – news, sport, current affairs and events. A studio drama, for example, would be rehearsed and then transmitted – beginning at the beginning and going on to the end. If it was to be repeated then the whole production would be remounted in the studio at some later date. Any mistakes would be seen by the entire viewing audience. If something went drastically wrong, the only outlet was for the caption *NORMAL SERVICE WILL BE RESUMED AS SOON AS POSSIBLE* to be shown.

People still sigh for the 'golden' days of television. They say it lent an element of excitement and adventure which is altogether lacking from pre-recorded programmes and many actors feel that their performances were given a definite edge in the knowledge that they were being seen, live, by audiences of thousands – soon increasing to millions. There have been attempts to recreate live drama with mixed success. Television has moved on and both audiences and programmes have grown so sophisticated that attempts to turn back the clock are not really viable.

Film recording

In the early 1950s it was obvious that there was a great need for recording television programmes, and a system of tele-recording was devised. Known in the United States as a *kinescope*, it was made by setting up a specially designed 16 or 35 mm camera, with a magazine holding up to one hour of film, in front of a high quality TV monitor. The pictures were literally filmed off the screen.

Its advantages lay in that as a film it could go through the same post production in terms of editing and dubbing as a film and it was also valuable as a means of retaining live programmes which would otherwise have been lost.

Its disadvantages were that:

- it was expensive,
- the film recording had to go through the processing stage in the laboratories which was time-consuming, and
- on transmission through a telecine channel there was a marked loss of picture quality.

The advent of videotape

It was not until the late 1950s/early 1960s that the use of videotape began to revolutionize television although experiments in magnetic recording had been going on for years before that.

For the first time programmes could be stored on tape and, unlike film recording, could be transmitted and repeated at any time without significant loss of picture quality. A library could be organized and programmes kept – if not for posterity – for the optimum life of videotape is still unknown – at least for a considerable number of years.

But perhaps the greatest advantage of videotape lay in the speed with which one could put together a programme comprising many different source elements and get it on the air. This opened up far greater possibilities for programme making especially as the initially primitive editing facilities improved.

During the past few years the technological revolution in all areas of video with the lighter, more versatile cameras and the advances in complex computerized editing has provided fresh scope in programme making.

Television today

Television today can be immensely complex. The ingredients in any one programme can be obtained from a number of different sources and the PA needs an overall understanding of the processes involved and her own specific job in helping to weld the many components into a single production.

'To begin with, almost everything was transmitted live.'

3 Multi- and single-camera shooting

Since this book will talk about both multi- and single-camera shooting, it might be helpful at this point to define just what is meant by these terms and how they originated.

A theatre or concert-goer will see the entire performance from one angle only, from the seat he is occupying in relation to the stage. He may have a good, clear, unobstructed view of the stage or his enjoyment may be marred by having to crick his neck at an awkward angle because some object is obscuring his vision. At all events, unless he gets up and moves from his seat to an unoccupied one elsewhere, he will have only one view of the stage and a wide angle one at that.

But while that is perfectly acceptable in the theatre and at concerts it would prove very boring if television were shown from one angle only.

This discovery is not, of course, new to television. Years before, the development of film gave rise to a new art form, whereby through the juxtaposition of shots of different size and angle, creatively edited together to form a smooth-flowing composite whole, the audience is guided through the story, seeing it, in a sense, through the eyes of the director's interpretation of the script. In order to achieve that effect the technique of out-of-sequence shooting was employed, using just one camera.

Single-camera shooting

Shooting out of sequence using only one camera means that the script is not generally shot in strict chronological order. Scenes are shot according to factors such as the availability of actors, different locations, ease of setting up etc., all determined by over-riding considerations such as the budget and speed of shooting.

But not only are scenes shot out of order. Because only one camera is involved, shots within each scene are also taken out of order, the entire scene, or parts of it, being repeated for the camera to record different angles and shot sizes.

This entire mass of disjointed shots, rather resembling the pieces of a jigsaw puzzle when the box is opened, are then assembled by an editor in the correct story order with the individual shots creatively joined together in order to give dramatic meaning and pace to the programme or film.

This method of shooting allows a great deal of flexibility at all stages of the production. Although any good director will plan his shots with an ultimate concept of the finished work in mind, there is room for experimenting, for change both during filming and, most importantly, in the editing stage.

Multi-camera shooting

Because television was originally live, it was, by its very nature, unable to adopt the film technique out-of-sequence shooting because there was no possibility of post production work before the programme was transmitted.

So there remained the alternative of transmitting a programme which used a single camera rather in the manner of a member of the audience at the theatre seeing the programme from one angle and one shot size only, or evolving a system whereby the viewer would see the programme through a series of differing shots and angles, as they would a film, but which could be transmitted live.

Multi-camera shooting.

The answer was, of course, to use a number of cameras simultaneously. Each camera would be positioned in the studio to give shots from pre-arranged angles. The work of each camera would be plotted beforehand by the director on to a camera script and in the control room of the studio the vision mixer would cut instantaneously from one camera to another in order, according to a pre-determined script. This would straightway give a succession of shots of different size and angle while showing the scene played through in story order.

Advantages and disadvantages

The technique of multi-camera shooting has its advantages and disadvantages like everything else. One advantage is that actors and performers are able to give more consistent performances than with single-camera shooting. In addition, the days spent in the studio produce a more or less finished entity.

Its main disadvantage, however, lies in the rigidity of the system in which everything has to be pre-planned to a degree that some directors find unacceptable as it allows very little flexibility when recording the programme.

When videotape came into common use in television it was possible for post-production editing to take place. But because, in its infancy, videotape editing was cumbersome, if not primitive, the multi-camera system was widely used for all manner of television programmes with only the minimum of editing afterwards.

Nowadays, with the increased sophistication of videotape editing, it is possible to combine some of the freedom of the film-style technique while recording multi-camera. For example, a drama shot multi-camera will probably record scenes out of order and the scenes themselves might be shot more than once, perhaps to allow for additional camera angles.

The need for flexibility

This means that the PA needs to have a greater flexibility of approach than hitherto. It is more than likely that studio productions will be recorded in whole or in part with one camera only and for the sections shot multi-camera to be recorded out of sequence. This necessitates a degree of continuity to be observed on the part of the PA together with accurate recording notes compiled for the editing.

4 Programme types

After differentiating between multi- and single-camera shooting, we should perhaps look at the different types of multi-camera programmes that are made.

Programmes fall into roughly three groups: ones which are pre-recorded, ones which are recorded 'as-live' and ones which are transmitted live.

Pre-recorded

The majority of multi-camera programmes fall into this category. These programmes are recorded with the intention of being edited afterwards. This means, among other things, that there is no necessity to record the material in the final programme order.

The PA is likely to be very involved in the setting up and content of the programme, and her role during the studio recording is one of keeping the studio informed about what is happening and making extensive notes for editing so that she can locate the relevant source material easily.

Some of these programmes might be closely scripted, of which drama is the obvious example. The PA on a drama would go into production knowing exactly what the programme content is, right down to what is going to be said and when. The camera positions and angles are generally fairly well worked out in advance and there is not a great deal of variance during the studio recording.

But pre-recorded programmes might also be unscripted, or only loosely scripted, with musical items and sections that are 'as directed'. Examples include: chat shows, game shows, studio discussions, interviews, consumer programmes, magazine programmes.

The aim during the studio recording would be to record these sections in out-of-order chunks and then edit them afterwards in the final programme order. Any previously recorded and edited inserts would be included during post production, as would any graphics, the opening and closing titles, name supers and so on.

As-live

If the intention is to pre-record the programme but treat it 'as-live' then it generally means that, as with 'live' programmes, the PA's first responsibility is one of timing. She will need to keep notes for editing provided that some editing has been allocated in case of problems, but the intention would be for the programme to be recorded in its final order to a fixed overall duration.

These programmes might contain all the elements – closely scripted sections, 'as directed' sections, music, pre-recorded inserts etc. – that have been mentioned above, but the difference for the PA will be in the priority she gives to timing during the recording.

Programmes often recorded 'as-live' could include sports events, music events (concerts, opera etc.), chat or game shows with a studio audience, studio discussions or magazine programmes.

In a programme with interviews and discussion, the intention might be to record it in the correct order, with VT inserts being rolled in at the right moments, but deliberately to allow the interviews or discussions to over-run in order that the best bits can be selected in subsequent editing. In that instance, the PA's timing role is of less importance than her notes for editing.

Live

Live programmes, such as the news and topical magazine programmes, clearly cannot be edited before transmission. Although they might be planned to a certain extent in the run-up to the programme, they are totally flexible in that they can, and often are, changed while on air.

The PA working on these programmes will have little or no involvement in setting up or programme content. Her overriding responsibility is one of timing to ensure that the programme gets on the air at the appointed time, comes off the air in a controlled fashion and that the programme flows smoothly from one section to the next.

Other live programmes might include sport, music, and national and international events. These could range from events such as the Chelsea Flower Show, the General Election, a 'Telethon' or 'Children in Need' appeal, the Olympic games or a world-wide pop concert in aid of charity. Often these programmes would be outside broadcasts or a mixture of OBs, studio and satellite. Depending on whether the programme must adhere rigidly to the permitted time slot or whether it is permitted to be 'open-ended', the PA might have priorities other than timing.

5 Tools of the trade

If you are starting out on your career as a PA, about to dip your toe into the icy waters of the freelance world, what are the essential tools of the trade?

Typewriter

Office equipment and stationery will no doubt be provided with the job, but a portable typewriter is essential, especially for location work. You will probably use a word processor in the office and some companies have invested in software packages such as Unilink's Scriptmaster which are designed specifically for television. If you work on a news programme you will probably need to master Basys or one of the other computerized systems.

Stopwatch

The badge of office of any PA has got to be the stopwatch, but what is the best all-purpose watch to buy? This will depend on the type of work you do.

Studio

If you can only afford one watch, make it a good quality analogue one with a clear minute hand. You will find that the option to flyback to zero on the side button is helpful. Get a watch that is tough, so that it will survive being dropped.

If you are involved in live or as-live programmes, then a second watch for use as a master overall watch is essential. This could be another analogue watch or a digital one. Seiko brought out a stopwatch called the Soundproducer that calculated time, something invaluable for last-minute calculations and as a double-check on your sums. At the time of writing, however, Seiko have stopped production of this watch. I believe they are considering whether to bring out a replacement. If you buy a digital watch, make sure it does not emit an irritating 'beep', as you will upset everyone in the gallery.

Location

For location work you only need an inexpensive, lightweight watch, either digital or analogue. If digital, again make sure that it does not beep.

'She will carry around with her what seems like the entire contents of her office.'

Clipboard

The clipboard is an essential tool for the PA which might, or might not, come provided with the general office stationery. If not, get a solid old-fashioned clipboard with a large, firm clip on the top. Designer clipboards with jazzy motifs might look good but they are not very efficient.

Polaroid camera

This is more for location work, although it could come in useful in studio situations recorded out of sequence. It is not necessary to buy a top of the range model – which, incidentally, would cost a great deal of money – but do not buy the cheapest either. A camera that gives a clear picture in reasonable colour and which folds neatly away will be most useful.

Other than the above, the PA travels from job to job unencumbered by much equipment – unlike the times when she is on location or in the studio when she will carry around with her what seems like the entire contents of her office!

6 Who's who

A whole army of people are employed at different stages during the making of a television programme and any production is the result of the coming together of a whole range of creative talent.

The core of this army is the production team, comprising those people who will work on any particular programme. Their number can vary depending upon the requirements of the programme and the way the jobs are defined within the company. Their titles, too, vary from company to company, generally depending, in the UK, whether the core staff came originally from ITV, the BBC or the feature film industry. This can cause a great deal of confusion. In the list below, I have tried to give as many as possible of the alternative titles of those basically doing the same job.

Recorded programmes

Executive producer

The executive producer carries overall responsibility for a series of programmes, for example a drama series or serial.

Producer

Together with the executive producer, if there is one, the producer frequently commissions the scripts, might be responsible for engaging the director and is closely involved in casting. He/she has overall responsibility for bringing the programme in within the budget allotted and time span agreed. The producer might have to work out the initial budget and allocation of facilities or that might be the job of the production associate.

Production associate

The production associate is responsible for the overall budget and for booking basic facilities, i.e. studios, OB crews etc. They also have the thankless task of trying to keep the rest of the production team within the confines of both budget and time scale allocated.

Writer

The writer is not technically part of the production team but it would seem churlish to leave him or her off our list of key personnel! The writer's work will, hopefully, be finished by the time the production team swing into action, but all too often the writing and re-writing is still taking place well into the setting up period to the frustration of everyone and the added workload of the PA. It is not all the writer's fault, however, for ensuring that the rehearsal scripts are finished in good time is clearly the responsibility of the story editor.

Story editor (also known as script editor)

Sometimes the producer takes on this function but frequently there is a separate story editor whose job it is to guide the writer through the hazards and pitfalls of writing for television. They need to provide both support and criticism where necessary, based on their expert knowledge of the television medium, and ensure that the writing is to the specified time slot.

Director

The director's responsibility is to the script. He/she provides the creative interpretation of the script, turning the printed page into the pictures and words that form a television production. The director must be a creative person, sensitive to the writer's intentions whilst exercising his own craft. He must have a good understanding of the mechanics of television production. He must be able to inspire actors as well as the production team with his image of the finished work. He must be able to communicate ideas and be receptive to the ideas of others. The director must be able to weld together the disparate talents of his team to form a harmonious whole. The director should inspire confidence as well as exhibiting all the qualities of leadership including the single-minded ruthlessness to get the job done.

And if the above paragraph does not equate with any directors that you know personally, remember that we all fall very far short of the ideal!

Directors vary in sex, shape, age, temperament and experience. There is no typical director just as there is no typical PA or anyone else.

Because the PA works very close to and with the director – in some respects the job could well be termed 'director's assistant' rather than 'production assistant' – this liaison can be fruitful and very much to the benefit of the production as a whole, or it can be uneasy if not damaging.

Scenic designer

The designer is responsible for the overall design of the set, whether studio or on location and will work in conjunction with the director to create the mood and atmosphere of the production.

Production manager (also known as the first assistant or floor manager)

The production manager joins the production some time before rehearsals begin. He or she will work closely with the director and actors, directing extras, issuing artists' call times and working out the recording order on studio days. The PM will be in overall charge of the studio floor. Sometimes the PM will work out the schedule for outside broadcast recording, sometimes the production assistant will do this job, but they will work in close liaison. The production manager will normally go filming or out on OBs.

Floor manager

The job of production manager is sometimes split, with the floor manager being solely responsible for the studio floor during the recording or transmission.

Location manager

The location manager finds venues for location work, whether film or OB, negotiates payments for the use of locations and is generally in charge of this aspect during the shooting.

Stage manager (also known as 2nd assistant or assistant floor manager)

The stage manager is responsible for any action props, for prompting artists, for cueing when required, for updating the script and for continuity in the studio.

Floor assistant (also known as 3rd assistant, call boy or runner)

The floor assistant works directly to the floor manager in calling the artists when needed and carrying out any other function assigned by the FM.

Casting director

The casting director has a specialized knowledge of actors and will work closely with the director in casting the production. The casting director will suggest possible actors for the different parts, will find out their availability, will arrange auditions and will book the actors. If there is no casting director this job is done by the director, assisted on occasion by the production manager together with the PA.

Costume supervisor

The costume supervisor is responsible for the design of the artists' costumes. The costume supervisor might have one or more assistants during the setting-up period and dressers will be assigned to the show on production days.

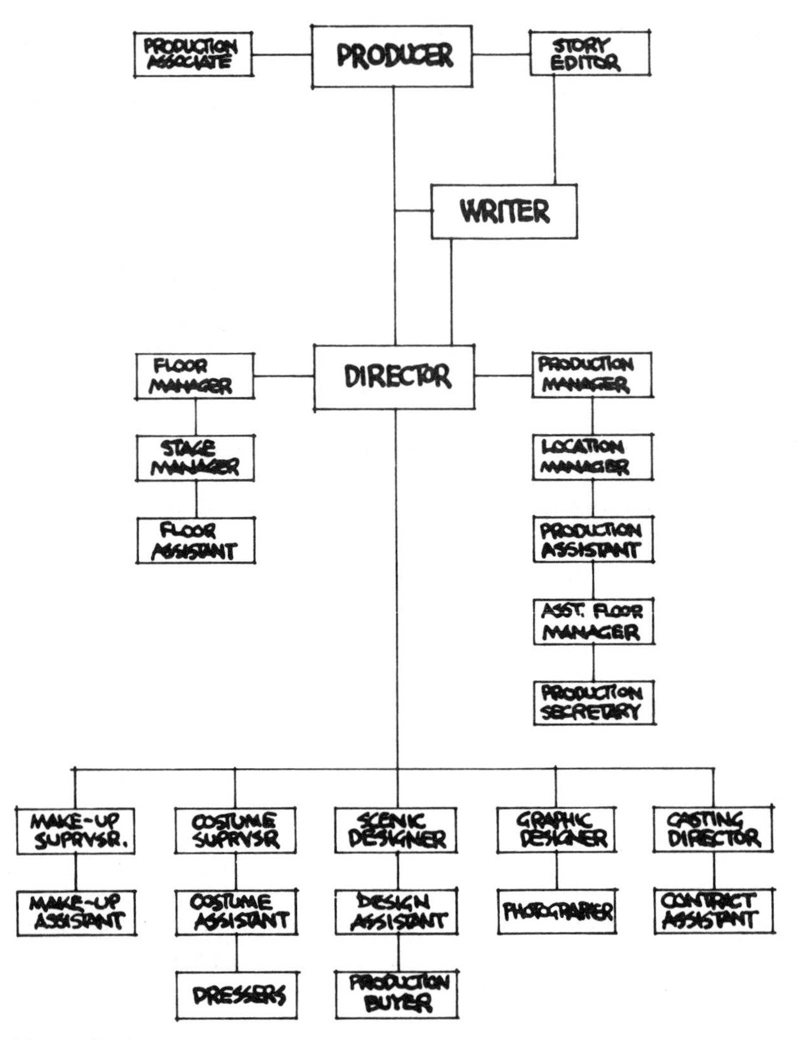

The production team.

Make-up artist

The make-up artist is responsible for the artists' make-up and will have one or more assistants during production days.

Graphic designer

The graphic designer will be responsible for opening and closing titles and any other art work for the production.

Production buyer (also known as properties buyer)

The production buyer is responsible for the hire or purchase of set dressing and will work closely with the set designer.

Production co-ordinator

The production co-ordinator does the same job as a PA in setting up and clearing up a production, but does not generally have any specialized studio or location skills. Sometimes a PA might be called a production co-ordinator. The name originates from feature films where the job of PA is effectively split between that of production co-ordinator and script supervisor.

Production secretary (also known as producer's secretary)

A production secretary's work will often overlap with that of the PA. Frequently they will work to the producer, typing the rehearsal scripts, budgets etc. and have much to do with the early preliminary work of setting up the production.

Live and news-type programmes

The structure and personnel involved of these types of programmes can differ from the above.

If one wanders into a television newsroom it looks, in layout, very like the busy newsroom of any newspaper with its telex and printout machines from news agencies, its banks of telephones, computers and newspapers, together with the constant chatter and bustle of a room full of busy people. The main difference being that in a television newsroom there is at least one, if not more, television set mounted in a prominent place, permanently switched on even if the sound is turned down.

It can appear chaotic and disorganised – the kind of place where 'I need it *now*' means that it is an hour overdue – and where everyone looks as if they are living on their nerves, their gastric ulcers and their stale, two-week old sandwiches, the smell of which permeates the room. It can, conversely, appear organized and ordered, but appearances are, in any event, deceptive.

Leaving aside the environment therefore, let us look at the people:

Producer of the day

Under the overall control of the head of the department or the series editor comes the producer of the day (who might also be called programme editor). He or she is basically responsible for the programme content, for the placing of each item in the running order and for the length of time allotted to it. The producer, under the overall department head, has the final say on editorial content.

There might, in addition, be a chief sub-editor, whose basic job is to vet the journalists' reports.

Reporters/journalists/researchers

There will be a number of reporters/journalists/researchers. Their brief is to cover whatever is newsworthy, research into it, report on it and present their item in packaged form for the programme. Their sphere of operations covers a wide range.

Programme director

The programme director's job is to weld together the many components of the programme into a whole during transmission. There will probably be a number of directors who will work on a rota basis.

Station assistant/news operative

Some companies employ station assistants or news operatives who might be required to undertake a variety of jobs: research, direct insert items, co-ordinate graphics, vision mix or floor manage.

Organizers/clerks

The terms 'organizer' or 'clerk' is only used in some companies. In others this function is covered by people as diverse as secretaries and station assistants and their duties may range from booking any kind of facility that is required – studios, video and telecine channels, graphics or post office lines – to arranging the travel details of reporters and crews, or obtaining, sorting and mounting stills and so on.

It is true to say that many of the jobs that would be the responsibility of the PA in other areas of television production are, in the live news-type programme, given to others. This is because of the special demands the live gallery makes upon the PA. The heavier the demands in terms of the amount and accuracy of the timing, the less involvement the PA has with any aspect of the programme content.

Teleprompt operator

The teleprompt operator is the person who types the scripts for whatever teleprompt system is used in the studio. In some companies where computers are used in the newsroom, the job is occasionally done by a PA. Autocue, Portaprompt, Digiprompt, are all trade names for the device which is mounted

The organization of the newsroom.

on to a studio camera enabling the presenter to read an enlarged image of the script while apparently looking directly at the viewer. The term 'teleprompt' will be used throughout this book.

Production assistant

The number of production assistants working on any particular programme of this type can vary from one to six or more, with a possible back-up team of typists and secretaries. The rota of work likewise varies. If the programme goes out nightly, there might be a total of three PAs alternating the work on a rota basis. One would do the gallery, one would act as co-ordinator and runner for the gallery PA and the third would clear up last night's work, sort out costings, music details, contracts, invoices and so on. The PAs tend to work as a team, assisting one another at times of intense pressure.

Typists/secretaries

Might be involved in general office work, they might book facilities and anything else needed. Their work frequently overlaps that of the PA and the organizers/clerks in the field of pure administration.

Basic studio personnel

The studio can be divided into two parts, the studio floor and the control room, which is often known as the 'gallery', the term originating from the very first studio in the world at Alexandra Palace where the control room was housed in the old minstrel's gallery.

Studio floor

Safety and discipline is under the control of the production or floor manager. Working under him/her is the stage manager and the floor assistant.

The scenic designer and assistants will be supervising the work of the set dressers, scene shifters, painters and carpenters.

The technical operations crew can be subdivided into cameras and sound. The camera crew work under the senior camera operator and the boom swingers work under the sound supervisor. There are also a number of electricians.

A teleprompt operator might be either on the studio floor or in the gallery.

Costume and make-up assistants will also be available if required to effect running repairs.

Gallery

The gallery might be divided into three distinct rooms, production control, lighting and sound. In a small studio, these may be contained in one area.

Production control

Apart from the producer, director and PA, there will be the studio supervisor (also known as the technical co-ordinator or technical manager in this area). The studio supervisor is overall in charge of the technical adminstration of the studio.

Unless the director does his own vision mixing, the vision mixer will sit beside the director. The vision mixer's job is to switch between the output of the different cameras and control the electronic effects. For programmes with complex digital video effects there may also be a video effects supervisor or second vision mixer.

If required there will be an operator of the caption generator, who will generate name superimpositions.

Lighting

In the lighting gallery will be the lighting director. Under him will be the vision controller (if there is one), who matches and adjusts the quality of the pictures from the cameras.

Sound gallery

In the sound gallery will be the sound supervisor who will mix and monitor the sound element of the programme. He will be aided by sound assistants.

Part Two: Setting Up

1 In the beginning . . .

The work involved in setting up a production will differ to a certain extent, depending upon the type of programme to be made. There are, however, some aspects of the work that are common to all kinds of programmes.

Paperwork

One of your main jobs in this setting up period will be to act as central co-ordinator for the programme. A programme amasses quite an amazing amount of paperwork, some of it generated by yourself as you book various facilities, some of it by others associated with the production.

'. . . the average television production marches along a road paved with memos, requisitions and paperwork in general.'

If an army is reputed to march on its stomach then the average television production marches along a road paved with memos, requisitions and paperwork in general. This is not true of all companies or all programmes but there is usually a vast number of forms, the greater part generated by the PA.

Computers using word processing software can save much of the repetitious drudge of typing and re-typing, but it does nothing to abolish much of the paper that will clog up the desks and filing cabinets of your production office.

Stationery

You should ensure that you have adequate supplies of stationery.

Programme file

As it is essential that all this paperwork be kept and filed in an easily retrievable fashion one of your first jobs should be to open a programme file – the bible of the production.

Address book

You will also need an address book with the names, addresses and telephone numbers of all those working on the production, together with anyone associated with the production.

Check list

Compile a check list of what you need to do when. You will find it a very useful aide memoire.

Distribution list

The company you work for might well have a distribution list, i.e. a list of the people to whom you need to send scripts, memos, booking forms and so on. As these vary from company to company it is not possible to give any standard of reference.

All this is, of course just good secretarial practice but one of your functions as PA is to be a good secretary and it is not an aspect of the job to be despised.

Budgeting

All programmes require a budget, and while the PA is generally not involved in working out the overall programme budget, she will very often have the responsibility of the day-to-day running costs, ensuring that these remain within the agreed parameters.

Booking

Facilities

The PA will be required to book the facilities needed by the programme and the requirements needed will be many and varied.

In a large company, the PA would book these facilities through an appropriate department, filling in booking forms as required. In a small company the PA might well be negotiating directly with facilities companies and specialist agencies. If this is the case, it is important to find out from the director precisely what facilities are required. If the director does not know exactly what is needed, do discuss the technical aspects of the booking with someone who is an expert in that area. A lot of money and time can be saved that way.

The largest facilities you might have to book will be the studios in which to record your programme and the editing facilities for post production. You must ascertain whether these are to be 'dry hired', in other words without any technical staff, or do they come with their own studio staff, videotape operators, editors etc? You need to find out precisely what you get for the hire and whether it covers all your requirements, or whether you would need to hire extra equipment or facilities for all or part of the time.

Always confirm a booking with a letter giving the precise nature of the facility booked, the agreed fee, as well as the dates and times required.

Crews/technical staff

Do try to think of contingencies. For example, does the crew provide the videotape stock or do you need to buy that separately?

Technical requirements

The technical requirements will vary from purchasing videotape for the recording and post production, hiring a teleprompt machine and operator, a caption generator and operator, cameras, sound equipment to visual effects etc.

Costume, props and drapes

For a production of any size, you will have production staff specifically dealing with these items, but on a small production the PA might be required to hire or buy them direct.

Artists

If there is a casting director, the PA will have little to do with the selection of actors, auditions and so on. If there is no casting director the PA will probably be required to assist the director in checking artists' availability, in arranging auditions – making sure that scripts are available – and sometimes in attending the auditions.

Once the director has selected his cast, the PA will ensure that they are contracted, either by sending a form to the relevant artists' booking department if they work for a large company, or arranging the contract after discussion with the director and artists' agent.

The information for the contract should be as detailed as possible and cover the following points:

- The character which the artist is required to play.
- The overall length of the engagement.
- The number of episodes in which the artist will appear (if a series).
- The studio/location shooting dates and times.
- Rehearsal dates and times.
- In the case of a series, whether the recordings will be one episode at a time or multi-episodic, i.e. on one studio day scenes from a number of different episodes to be recorded.
- Rates for overtime (generally as laid down by Union agreement) unless an 'all in' deal is agreed.
- Any special clauses, i.e. that hair should/should not be cut, that a beard or moustache should be grown.
- If the programme is to be a co-production and the name and address of the co-producer(s).
- The use to be made of the performance, i.e. does the contract cover sales to other countries and if so, which? Is the artist contracted for a single transmission or does it include a repeat? Has provision been made for home video sales?

Once the contract has been issued a copy will be sent to the PA who should read it through to ensure that it is accurate on all points.

Stunt artists

In addition to the above information, the PA should include the nature of the stunt that is to be performed.

Supporting artists and walk-ons

Supporting artists and walk-ons are non-speaking parts or characters who have a few unscripted words or lines to say. Working from the director's requirements, the PA will need to state the number required, dates, times and places, whether male or female, the approximate ages and type, i.e. farm labourers, assorted office workers and so on, when booking supporting artists and walk-ons.

Children

There are strict laws in the UK governing the employment of children, i.e. those under the age of 16. These should be read and understood by any production which needs children in their cast and the schedules worked out accordingly. It will be necessary to engage chaperones and tutors and the children's hours of work and study should be strictly regulated. It is generally the PA's job to keep a note of the hours for submission to the inspector.

Foreign artists

Foreign artists (those from non-EC countries) will require a permit to work in the UK and the nature of the work must be specifically stated. The Department of Employment requires a minimum of six weeks to process permit applications.

Presenters

Many presenters on magazine-type programmes, news and current affairs are employed on a long-term contract. If they wish to take other short assignments when under contract, they must seek permission.

Musicians

The rules and regulations governing musicians are extremely complicated. There are strict rules relating to rehearsal times, doubling, performance schedules, overtime etc. The PA needs to tread very cautiously when it comes to booking musicians. It would not be appreciated if she filled the studio with a symphony orchestra when all the direcor wanted was a small string ensemble! Do take advice from your composer or musical director.

The PA must seek guidance when booking artists as any mistakes could be extremely costly and even jeopardize the entire production.

2 Wall charts, meetings

'Which came first, the chicken or the egg?' is an old saying and one which could be accurately applied to almost any production in the setting up period.

Some people would assert that no constructive work can be done before the scripts have been finalized. But anyone who has worked in production will know that work often starts on a programme without the final scripts – sometimes without any scripts at all, purely an outline.

It can be difficult to assess the number of days' rehearsal and shooting required until the scripts are in a finished form and therefore difficult to schedule the weeks ahead, but it may well be that a studio has had to be booked a long time in advance. It will then be more a question of fitting the scripts into the pre-allocated facilities rather than tailoring the facilities to fit the scripts.

Whatever the situation pertaining to your production, you should, as in all PA work, do what you can when you can, and one thing you can do fairly early on is provide wall charts.

Wall charts

Day by day

First of all a chart showing each day of the production should be made. This chart can be manufactured using different coloured cards – each card representing one day and the different colours clearly showing rehearsal days, studio days and so on or you could use coloured pens on a whiteboard. If your company has specially designed charts you should use them.

Any information relevant to the production should be written on the chart and kept up to date daily. Meetings, recces, travel days, days off . . . everything should be shown.

Production list

The names, addresses and telephone numbers of the production team should be written in large letters on another chart.

Artists' chart

The movement of artists – whether travelling, rehearsing, on location, in the studio, costume fittings, make-up, photo-calls, etc. can, with a large cast,

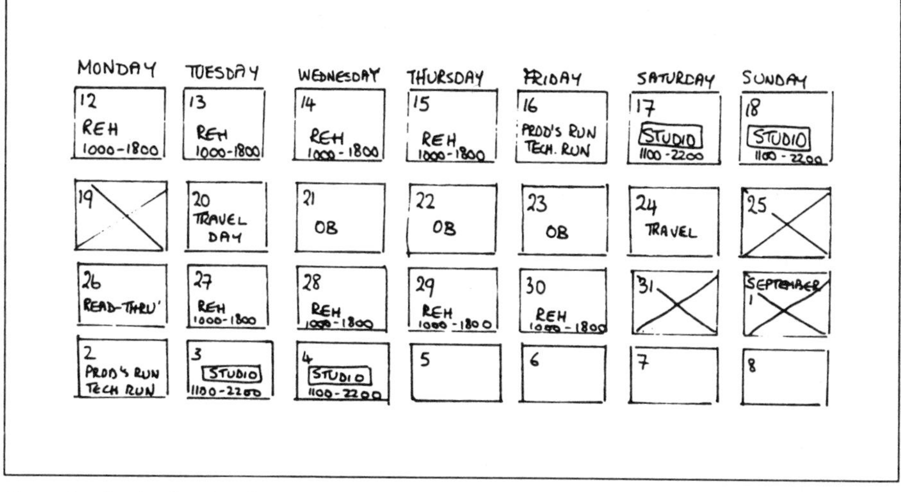

Example of a day-by-day chart.

become extremely complicated. The only way to keep track of them is by another chart which will show you at a glance who is doing what, where, on each day and will prove invaluable in working out supplementary payments.

Meetings

Before very much work can be done on a production, it is necessary to call together the key people involved.

ARTISTE	NO. EPISODES	AUG 12	13	14	15	16	17	18	19	20	21
FRED BLOGGS (JOE)	1, 3, 4 6	REH.					ST.	ST.	X	X	X
MAVIS LOW (HELEN)	1, 2 4, 5, 6	REH.					ST.	ST.	X	TRAVEL	OB
ARTHUR STEELE (MR HOW)	2, 3, 5, 6	TRAVEL	COSTUME FITTING	TRAVEL	X	X	X	X	X	TRAVEL	OB
JOHN CARLISLE (STEPHEN)	1 5, 6	REH					ST.	ST.	X	X	X

Example of an artists' chart.

Studio

Once a date has been set for the planning meeting, the PA will notify all members of the production team together with the senior technical and craft staff.

The designer would normally bring scale models of the studio sets to the meeting together with the floor plan to enable everyone to see exactly what will be involved.

The director will then talk through the production and discuss all aspects: the technical requirements, cameras, sound, special effects; costume and make-up requirements; design; the recording schedule and so on. Any problems will be discussed and fully thrashed out.

During this meeting the PA will take notes and circulate the decisions reached to everyone who attended.

'The director will talk through the production ...'

Location shoot

A similar meeting will be held for the technical staff and production team either in the production office or possibly 'on site' if it is to be an outside broadcast. At this meeting the location manager will outline the schedule and all the technical requirements will be itemized.

News-type programmes

An initial meeting will be held between the producer of the day, the director of the programme and the journalists. This meeting will be to throw ideas around, to check work in progress and list who does what. If pre-recordings are to be made, possibly an interview or the putting together of some composite package, this would be noted by the director. The PA might or might not attend this meeting.

At the planning meeting the draft running order is given out by the producer. He or she will list the items, state, as far as is known at this stage, the technical requirements needed, the source from which the items are to be generated, the placing of the items in the programme – although this, as everything, may well change – and the length of time the producer has, provisionally, allotted to each item. Any information about the programme will be given to the meeting and the journalists will report on the stages their items have reached. From this information the PA will add up a rough overall timing and circulate a draft running order.

3 The cast

Cast list

Once the cast has been finalized and booked, the PA should type out two lists, one a simple cast list for general distribution, the other a more detailed list with addresses, telephone numbers, agents' names and telephone numbers, the episodes in which the actors will appear and the dates for which they have been contracted. This list should be restricted to the relevant members of the production team, i.e. the producer, floor manager, costume and make-up supervisors.

Arrangements for rehearsal

You will need to book a suitable venue for rehearsals and, if necessary organize transport. You will also need to make arrangements for catering.

Arrangements for recording

Once the studio schedule is worked out it will be necessary for the PA to book dressing rooms for the cast, arrange catering for the recording days and possibly organize accommodation and transport.

Information to cast

The PA will then write to the actors with the following details:

- scripts,
- time and place of the read-through,
- times and places of:
  –rehearsals,
  –location shooting (sending a schedule if it exists at that stage),
  –studio dates,
- maps as necessary to get to the rehearsal room/location/studio,
- details of accommodation if appropriate.

Time sheets/studio call sheets

Either you or the floor manager will keep a note of the hours and days worked by the artists and extras. You must ensure that supplementary payments for additional hours and days worked are paid.

Photo calls

The PA might be required to book photo calls – arranging the photographer and the venue and ensuring that the actors are informed and any arrangements made for travel where necessary.

'The PA might have been involved in getting together the audience.'

Studio audience

If there is to be an audience in the studio, the PA will need to inform reception, the commissionaires, the security staff and the studio operatives. The PA might have been involved in getting together the audience: by advertisements in the press, by contacting interested organizations or by other means. Alternatively this job might be done by another person who will then need to know the type and size of audience required.

4 Research, insurance, copyright

Research

This might or might not be part of the PA's responsibilities but it can be a very rewarding part of the job. You might find all you need to know from local libraries, but other places that might prove helpful are:

- town halls,
- local councils,
- government departments,
- law courts,
- welfare institutions,
- Army/Navy/Air Force,
- museums,
- universities and polytechnics,
- newspapers (local and national),
- film, video and newspaper libraries,

Insurance

If you are working for a small company, it might be your responsibility to arrange suitable insurance for the production. Go to one of the insurance agencies who specialize in film and television work and take their advice. The types of insurance you might need would be:

- public liability,
- employers' liability,
- special cover for artists,
- insurance to cover a re-shoot in the event of fire, accident, theft, loss or damage to the recorded tapes or shot film.

Copyright

Copyright laws are designed to give protection to the person, or organization, who has created the work and these laws are rightly strict. In 1989 the

Copyright Act 1956 in the UK was replaced by the Copyright, Designs and Patents Act 1988 which set out to restate the law of copyright.

The protection the law gives ensures that no-one other than the copyright owner has the right to use the work unless authorized by the owner, who is thus able to charge a fee.

It is particularly important that permission is obtained before any material is used in a production if that material is copyright.

Copyright in a literary, dramatic, artistic or musical work expires fifty years after the end of the calendar year in which the author dies. If the author is unknown, the copyright will expire fifty years after the work is first made available to the public by being performed, broadcast or exhibited. Sound recordings lose their copyright fifty years from the end of the year in which they were made, or from the year of their release. Copyright in broadcasts expire fifty years after the end of the year in which they were made and in cable programmes, fifty years after they are included in a cable programme.

In its simplest form the material could be a literary work, a painting, a photograph, a musical composition or a recorded musical performance. But such things as records, tapes, CDs, illustrated books, films and videos involve multiple copyright.

Music played live in the studio *must* be cleared for copyright purposes *before* the studio and any recorded music, particularly commercial recordings, must likewise be cleared. The type of clearance required should be checked first, i.e. for transmission in one country or world clearance. You will also need to know whether the music is 'visual', i.e. the actors can 'hear' it, or 'background', i.e. mood music to create an atmosphere which the actors on the production cannot 'hear'.

There are three main organizations in the UK who deal with the copyright use of music in a programme.

The Performing Right Society,
29–33 Berners Street,
London W1P 4AA.
Telephone: 071 580 5544.

This society provides for the collection of royalties for the public performance, broadcasting and diffusion of copyright music on behalf of its composer, lyricist and music publisher members and members of affiliated societies worldwide.

The Mechanical Copyright Protection Society,
Elgar House,
41 Streatham High Street,
London SW16 1ER.
Telephone: 081 769 4400.

This society protects the rights of composers and publishers of music by entering into agreements with users of music (record companies, film companies, broadcasting companies etc.) to receive royalties on behalf of its members.

Phonographic Performance Ltd.
14–22 Ganton Street,
London W1V 1LB.
Telephone: 071 437 0311/6

This organization covers the licensing of the public performance and broadcasting of sound recordings in the UK.

5 Schedules

Schedules are the creative effort of the immediate production team and are issued for everything: rehearsals, location shooting, studios. Sometimes they are worked out solely by the PA, sometimes by the floor manager (or production manager). The director, too, has a say in the schedule but the end product is the result of expertise in compilation and organization of the salient facts, placing them in a simple to understand and comprehensive order and that is the responsibility of the PA. On a complex programme the schedules can be works of art!

Rehearsal schedules

Normally the director and/or floor manager will work out the schedule, based on the amount of rehearsal the director wishes to allocate to each scene.

This schedule is usually simple and straightforward with dates, times, venues and artists involved, together with perhaps any special props needed as a reminder to the stage manager.

Outside broadcast schedules

Depending upon the length and complexity of the OB, this schedule could comprise many pages.

1. The cover page will contain the series title, the production number(s), the names, room and telephone numbers of the production team and details of the crew. It will also contain the distribution list, unless this is shown on a separate sheet.

Then, in no special order, there will be the following information:

2. Details of locations, dates, times, addresses, contacts and parking.
3. The actors involved in the location work, the dates required and episodes.
4. Technical information specific to an outside broadcast.
5. Accommodation details for cast and production.
6. Travel details for cast and production.
7. Catering information.
8. Useful contacts.

There will then be a comprehensive day-to-a-page breakdown of the shooting giving for each day:

- the location,
- the call times,
- the names, addresses and telephone numbers of the contacts for that day,
- catering arrangements,
- transport arrangements,
- directions to location,
- an hour by hour schedule of shooting,
- the order of shooting with scene breakdowns. The order will normally be worked out with reference to the following criteria:
  - the number of scenes at any one location,
  - the position of locations, bearing in mind travel needs,
  - the availability of locations,
  - the availability of actors,
  - the number of days allocated to the shoot.

At the back relevant maps will be attached with the locations circled.

Studio schedules

These form part of the camera script as the recording or running order. The schedule is worked out by the production manager in consultation with the studio supervisor.

Example:

TUESDAY 10 NOVEMBER 1992
1100–1300 Camera rehearsal
1300–1400 Lunch
1400–1530 Camera rehearsal
1530–1600 Line up (tea trolley ordered 1530)
1600–1800 Rehearse/RECORD
1800–1900 Supper
1900–2200 Rehearse/RECORD

Technical facilities

Another page should contain the technical facilities: the VTR channels and times booked, the cameras and any special facilities.

Cast list

There should be a page containing the cast list together with a list of dressing rooms allocated to the artists.

MONDAY 2 SEPTEMBER

LOCATION: THE OLD MILL HOUSE, Grangeley, Kent

CONTACT: Mr G. LEADER (Caretaker) Tel: Grangeley 465

LOCATION CATERERS: Fosley & Cole (Tel: 789 1234)

UNIT TRANSPORT: As Location Manager's sheet

DIRECTIONS: On entering Grangeley from the north, take the 2nd left at the roundabout into Lowescroft Road. Follow road to T-junction. Turn right into Shadley Road. 3rd left into Grangeley Avenue and the Old Mill House is immediately on your right.

CALL TIMES: 0800 at Lowescroft Hotel (for costume and make-up):

HELEN DEWSBURY (Joan)
CHRISTINE SHALTON (Harriet)
SUSANNA SHAW (Sarah)
ROGER SHINER (Harry)
PETER JOHNS (Graveney)

0915 R.V. ON LOCATION : Crew and artistes

1400 JOANNA STUART (Vicky) - driver to collect from hotel)

SCHEDULE:
0930-1300 SHOOT I/6, I/9
1300-1400 LUNCH
1400-1715 SHOOT III/1, IV/7
1715 WRAP

SHOOTING ORDER:

Ep/Scene	Pages	Int/Ext Time	Location	Characters	Cameras	Shots
I/6	1-4	Ext/Day	Garden by stream	Joan Harry Graveney	1, 2, 3	1-12
I/9	5	"	"	Harriet/Sarah	1, 2	13-18
III/1	6-9	"	Garden by mill wheel	Harry/Graveney Joan	1, 2, 3	19-29
IV/7	10-11	"	"	Vicky/Harriet	1, 2	30-36

Example of a schedule (day to a page).
NB The script page numbers will have been over-written to be consecutive for the shooting for the sake of simplicity.

Page No.	Shot Nos.	Ep/Scene/Set/Characters	Cam/Sound/Sp. reqs.	Synopsis/time of day
1-8 (4)	1-7	1.3 DINING ROOM, MANNERS HOUSE ALICE JOHN JAMES WILCOX MARY WILCOX ELWYN BRAND GRANT TYLER SARAH GLYN 2 waiters	1A/B/C A1/2 2A/B B1 SWINGER OPEN	Dinner party. Alice finds out about Tew & is regarded with suspicion by James. EVENING.
9-12 (24)	45-46	1.15 DINING ROOM ALICE JOHN	1C 2A 2B B1 SWINGER CLOSED	John accuses Alice of the robbery. MORNING
13-18 (61)	47-60	3.25 DINING ROOM SARAH GLYN JOHN ELWYN GRANT MARY WILCOX JAMES WILCOX DET. INSP. LANDER 2 PCs / 2 waiters	1A/B/C A1/2 2A/B B1 3A/B SWINGER OPEN	Det. Insp. and PCs interrupt dinner party to arrest John. EVENING.
		RECORDING BREAK	/1 TO D/2 TO C/	/COSTUME CHANGE JOHN/

Example of a studio recording order.

Photographer

If a photographer has been booked, his/her name, telephone number, the times he/she will be on the set and the artists required for a photo-call should be given.

Audience

The arrival time of the audience, if any, should be shown, together with their approximate numbers.

Studio recording order

Then we come to the pages which show the scenes listed in the order in which they are to be recorded. As you will see from the example on the preceding page the columns denote the following:

1. Script page numbers. Both the page number in *recording order* and the page numbers in their original *story order* are given (the story order page numbers are the ones in brackets).
2. Shot numbers. Each shot is numbered consecutively in the camera script. These numbers should be listed in the studio recording order.
3. Episode, scene, set and characters are listed next.
4. Camera/sound/special requirements. The information relevant to this column can be obtained from the director's camera script.
5. Synopsis/time of day. If there is room it is a good idea to include a brief synopsis of the scene.

Any recording breaks, camera moves, changes in sets, costume and make-up changes should be noted on the recording order. This information can be obtained from the camera script.

A draft recording order is often required for the technical run, but unless the camera script has been finalized before the run, the definitive recording order might contain a number of changes. This final recording order which is to be included in with the camera script is better typed *after* you have both typed and checked the main body of the script. There will be less likelihood of error that way.

The studio recording order is really an annotated form of the camera script. Many of the studio staff work solely from these sheets and additional copies should be made. They should be duplicated in a distinctive colour so that they stand out from the camera script.

6 Running orders

News-type programmes

Many newsrooms are now computerized, which has done away with the chore of constant retyping of scripts and running orders. Using a centrally linked computer, reporters write their stories directly on the computer. These stories are read by the producer and the presenter from the VDU and any changes made before it is printed out in script form. The running order is also compiled on the computer by the PA. It can be updated and changed throughout the day before being printed just before the programme. Some computers time the links and print the figure on the running order, but these should be re-checked by the PA as they can be inaccurate.

The running order

Each programme has its own format for setting out a running order, one that is easily understood and acceptable to everyone working on the programme. Some programmes like as much information as possible put down on the running order, others like the barest minimum. The example given over shows the following:

General details

The programme title, the date and programme number have been shown. The running time (R/T) together with the 'on air' and 'off air' times have also been typed in although these might be approximations at this stage.

The names of the key personnel have been shown, implying that these might vary from day to day.

Sequences

Every programme, whether live or pre-recorded is, at some stage or other, broken down into small, manageable segments. Because live programmes contain elements from many different sources it breaks down neatly into different sections that are placed in an order pre-determined by the producer of the day and linked by the presenter.

Each numbered sequence shown on the running order therefore consists of a separate story together with its link into or out of it. In the example below, the only exception is Sequence 5, where three short news items have been placed together under the overall title 'Shorts'.

DRAFT RUNNING ORDER WEEKDAY NEWS FRIDAY 27 NOVEMBER 1992

In : 18.00.00
R/T : 20.25
Out : 18.20.25

Prog. No: XYZ/1234/A

Producer : MARY JONES
DIRECTOR : DOUGAL McLEAN
P.A. : JOSIE BLANE
F.M. : FRED SMITH

Presenters : CHERRY STONE
ANDREW PLUM

SEQ.	SOURCE	TITLE/PRESENTER/REPORTER	EST. DUR.
1	VTR (A) STUDIO (Cams 1/2)	TITLES Cherry and Andrew	0.23 (0.30)
2	STUDIO (1/3) VTR (B) + Aston	Cherry links to LOST CHILDREN (PT)	(2.35)
3	STUDIO (2/4) VTR (C) + Aston	Andrew links to NEW TECHNOLOGY (TW)	(2.20)
4	STUDIO (1) VTR (A) + Aston	Cherry links to SOCCER VIOLENCE (LW)	(1.15)
5	STUDIO (1/2) VTR (B) (Mute) VTR (C) + Aston VTR (A) (Mute)	Andrew and Cherry SHORTS HOUSING PROJECT LOIS GREY BARN MURDER	(2.45)
6	VTR (B)	TEASE : HIDDEN TREASURE	(0.15)
7	STUDIO (2/3) VTR (C) + S/F's	Andrew links to WAR CRIMES (CP)	(2.10)
8	STUDIO (1/4) VTR (B)	Cherry links to STATE VISIT (LE)	(1.25)
9	STUDIO (2) VTR (A) + Aston	Andrew links to HIDDEN TREASURE (AB)	1.55
10	STUDIO (2/3) + Aston VTR (B)	Andrew links to SPORT	(2.35)
11	STUDIO (1) VTR (C)	Cherry links to WEATHER	(1.15)
12	STUDIO (1/2) + Aston	Cherry and Andrew CLOSING HEADLINES	(0.35)
13	STUDIO (1/2) VTR (A) + Roller	AND FINALLY . . . Cherry GOODNIGHT (C & A) PROGRAMME CLOSE	 (0.35) (0.15)

STANDBYS : RAIN FORESTS (TP) 2.10
INDUSTRIAL REVIVAL (CP) 1.10

Example of draft running order.

'Many newsrooms are now computerized.'

The final order may well change and some PAs leave gaps to allow for the addition of possible late items, thus spreading the thirteen items listed over numbers ranging from 1–20.

Source

The next column relates to the source from which the item originates.

Sequence 1 is a pre-recorded opening title sequence on videotape before coming to the studio to see the two presenters on cameras 1 and 2. The sources therefore for Sequence 1 would be VTR and studio.

Sequence 2 shows a studio link into the item 'Lost Children'. The cameras involved will be 1 and 3, camera 3 showing a still which will be overlaid on camera 1's shot of the presenter. 'Lost Children' comes from a VTR source and name superimpositions (Aston) will be timed in by the PA.

Sequence 3 sees another studio link into the item on 'New Technology'. Camera 2 will be giving us a shot of the presenter while camera 4 will overlay the relevant still. 'New Technology' comes from a VTR source and is now a composite package from an original film together with a pre-recorded studio interview. Name superimpositions are to be added.

Sequence 4 is a straightforward link into a VTR item.

Sequence 5, 'Shorts', consists of brief news items on VTR. The first and third items are mute and will require voice-over commentary by the studio presenters on cues given by the PA or floor manager.

Sequence 6 is a short 'tease', a preview of the item on 'Hidden Treasure' which comes later in the programme. The sequence is to be pre-recorded in the studio during the afternoon.

Sequence 7 is a studio link using cameras 2 and 3 into a VTR item with stills stored in a slide file.

Sequence 8, 9 and 10 are straightforward.

Sequence 11 is to be pre-recorded during the afternoon.

The closing headlines, Sequence 12, shows us the two presenters in the studio.

The closing sequence, 13, has a last comment and goodnights from the presenters on cameras 1 and 2 before going into the closing title sequence recorded on VTR.

Title/presenter/reporter

The next column gives the name of the story, the initials of the reporter responsible and the presenter who is to give the link for the item.

Durations

If no durations are known at this stage, this column could be left out. If there are only one or two known durations these could be typed in and other figures added as they are made available to the PA. Alternatively, this column could be used for the producer's estimated times. In the example above, the producer's estimated times are shown in brackets and the known times without brackets.

The durations for each sequence are, of course, made up of two, if not three, separate timings:

- the duration of the link into the item,
- the duration of the pre-recorded insert,
- a possible back link out of the item.

These can be shown separately and only added together in the later columns which the PA fills in.

Space could be left on the righthand side of the running order for the PA to work out the cumulative times, or this might be done on a separate sheet, a time chart, according to preference.

Standbys

The standbys are completed items which might need to be slotted in at the last minute if the programme is under-running or if another item is not ready in time.

OK	Seq:	Storyline	Source	Jrn Dur:	Total
		Date: **18.09.92** MIDLANDS TODAY 1/RMD/R037J TX time: 18:31'30"			
AXP	01	*******************HEADLINES	HERE+++++++++++ +++	0:21	
XXX	02	*SYMBOL+IDENT+TITLES+HEADLINES	LASER DISC + TXA	E:45	
	03			0:00	
	04	*BRIDGEWATER REACTION INTRO	DAVID	E:20	
XXX	05	*BRIDGEWATER REACTION/PWILSON	BETA+SUPERS	1:18	
	06			0:00	
	07			0:00	
	08			0:00	
	09	*BRIDGWATER LINK	DAVID+GFX	0:00	
XXX	10	*MALLOY EVIDENCE INTRO	SUE	0:14	
	11	*MALLOY EVIDENCE/PWILSON	BETA+SUPERS	E2:20	
XXX	12	*LIVE STUDIO INTERVIEW	DAVID+2	E2:00	
	13			0:00	
XXX	14	*CANNOCK ROBBERY INTRO	SUE	0:14	
XXX	15	*CANNOCK ROBBERY/MANNING	BETA+SUPERS	1:10	
	16			0:00	
XXX	17	*JAGUAR JOBS INTRO	DAVID	0:16	
XXX	18	*JAGUAR JOBS/FOSTER	BETA+SUPERS	E1:30	
	19			0:00	
XXX	20	*LITTLETON JOBS INTRO	SUE	0:18	
XXP	21	*LITTLETON JOBS/JEFFORD	BETA+SUPERS	E1:10	
	22			0:00	
	23			0:00	
	24	**************NEWS IN BRIEF***	******1.00******* ***	0:00	
AXP	25	*CHILD ABDUCTION	PRES + BETA OOV	0:19	
AXP	26	*BOBBY	WIPE TO BETA OOV	0:18	
AXP	27	*LIGHTNING	WIPE TO BETA OOV	0:19	
AXP	28	*SWANS (S/BY)	WIPE TO BETA OOV	0:14	
AXP	29	*SWAMI (S/BY)	PRES	0:22	

Example of running order produced from Basys.

7 Read-through/rehearsals

'The first time the entire cast is assembled together will be for the read-through.'

Read-through

The first time the entire cast are assembled together will be for the read-through. It is the PA's responsibility to book the room in which the read-through is to take place; to ensure that everyone has been notified of time and place and that scripts have been sent to the cast well in advance.

Read-through timings

The timings on a recorded programme are naturally more fluid than on a programme going out live, but they are important, nonetheless. During the read-through the PA should note down the timings in pencil on each page of script. She should also write down the running time of each scene and add them together cumulatively to give a rough overall running time. It should be stressed that this will only be an approximate timing, but if the slot allotted for the programme is fifty minutes and the read-through lasts one hour and ten minutes, she would be wise to inform the director.

In any case she will probably have a good idea already about the overall length of the scripts. If the rehearsal scripts are typed using the layout suggested in the next chapter then fifty such pages containing a mixture of action and dialogue will work out at roughly half an hour screen time.

Rehearsals

It would be unlikely for the PA to attend many of the rehearsals unless specifically requested by the director.

The stage manager and floor manager attend rehearsals and any script changes will be noted by the stage manager who will pass them on for re-writes and incorporation into the final camera script. The SM will also keep a note of the artists' hours and time the scenes more accurately.

Blocking

During the rehearsals the director will 'block' the scenes, working out both actors' moves and positions and the moves and positions of his cameras.

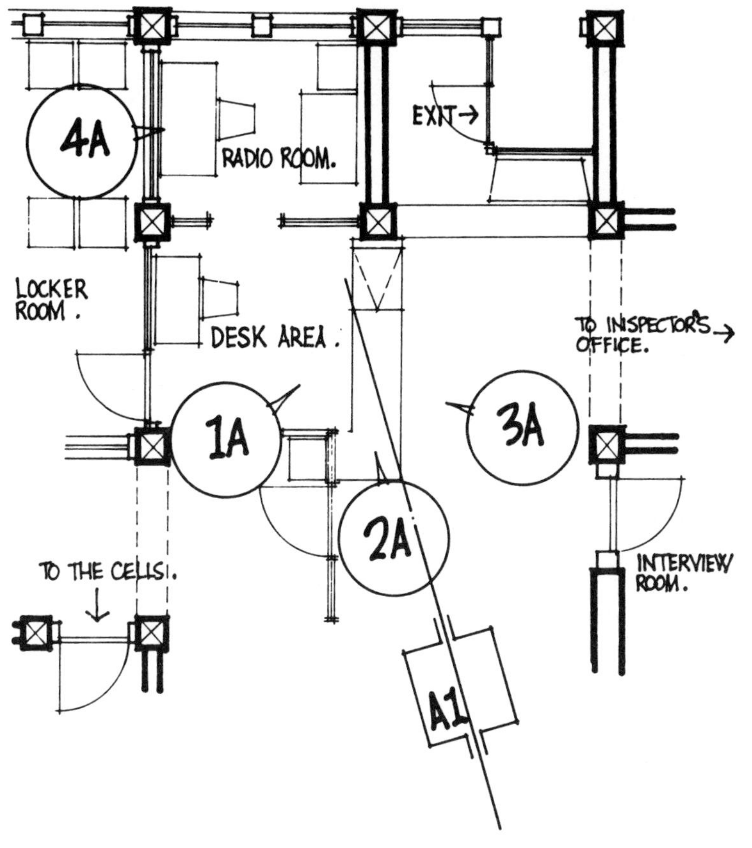

Example of a studio floor plan.

Floor plan

The camera positions will be plotted by the director on a studio floor plan, the sets having been drawn by the designer after discussions with the director. Copies of this floor plan will then be distributed to the studio staff.

Camera notes on script

During the course of the rehearsals the director will be marking the cutting points from one camera to another on his script and noting down the kind of shots he wants from each camera position. This will eventually form the basis of the camera script which is described in the next chapter.

Producer's run and technical run

The culmination of the rehearsal period is the producer's run and technical run when the scenes are acted in the order they will be recorded in the studio for the benefit of the production team and technical studio staff.

The director will point out camera positions and moves, lighting and sound and any problems will, hopefully, be resolved before the production goes into the studio.

8 Scripts

Rehearsal scripts

However the original manuscript is delivered – handwritten on the backs of old cigarette packets or typed out with painstaking care – it will be necessary for either the PA or the production secretary to re-type it in an acceptable form as a television rehearsal script.

Format

- Scripts should be typed on one side of the page only.
- The typing should be on the righthand side of the page to enable the director to plot the cameras on the left. It is not helpful to type stage directions on the lefthand side.
- Plenty of space should be given between the lines of dialogue, and stage directions should be easily distinguishable from the dialogue by typing them in capitals.
- Page numbering should be consecutive.
- Each scene should be started on a fresh page. This will make it easier for re-writes and amendments to be slipped into the body of the script and is *essential* for working out schedules.
- If you are working on a series, then in addition to the consecutive page numbering the PA should identify each page with its episode and scene number, i.e. Ep.2/Sc.3, or, more simply, II/3.

Scenes

Scripts should be likewise broken down into scenes, if the writer has not already done so, each scene being a different set or location. The scenes should be numbered and the heading should state whether an interior or exterior and the time of day, i.e.

SCENE 1: DAY. INT. CELLAR

SCENE 2: DAY. INT. HALL

Example of rehearsal script

– 10 – IV/4

SCENE 4: NIGHT. EXT. GARDEN (BY SUMMER-HOUSE)
THE SUMMERHOUSE IS ABLAZE WITH LIGHT, SENDING LONG SHADOWS ACROSS THE LAWN, TREES AND SHRUBS. SILHOUETTED AGAINST THE HOUSE IS LOUIS, WHO WATCHES ALICE MAKE HER WAY TO MEET HIM.

ALICE
I'm sorry . . . I couldn't come sooner . . .

LOUIS
I hardly expected you at all – after what's happened . . .

ALICE
So you know?

LOUIS
Yes. Here . . . sit down . . .

HE HELPS HER TO A SEAT

Colour coding

If you are working on a series then it is extremely helpful to employ a colour coding system. Each rehearsal script should be duplicated on different coloured paper, the same colours carried through the script breakdowns, schedules and so on.

Distribution

Whether or not you typed the rehearsal script, it is the PA's job to ensure that it is sent to all the relevant people – not forgetting the writer!

Your next job will then be to break the script down into a simple chart known as the breakdown.

Script breakdown

An example of a script breakdown is given below. Note that a synopsis of the scene is given. This will prove invaluable, especially when you are working on more than one script.

Camera scripts

In a multi-camera production, the director needs to have worked out the visual and sound coverage well in advance of the studio day. He needs to have assessed the number of cameras required, where they will be positioned

Page No.	Scene No.	Int/Ext	Day/Night	Location	Characters	Synopsis	Sp. requirements
1	1	Int	Day	Railway Stn.	Louis Alice Porter (W.On) Passengers	She meets him off train Arrange meeting in garden	F/P train
3	2	Ext	Day	Gravel pit	Ch.Insp. Bartley James Wilcox Policemen	Police find body	Body of Ashley Tew
4	3	Int	Night	Dining Room at Manners House	Alice John James Wilcox Mary Wilcox Elwyn Brand Grant Tyler Sarah Glyn	Dinner party where Alice finds out about Tew - is regarded with suspicion by James. Sarah and John exchange speaking glances!	
10	4	Ext	Night	Garden (by summerhouse)	Alice Louis	They meet. She tells him of her fears. He dismisses them.	

Example of a script breakdown.

initially, where they will move during the course of the action – taking care that he does not get their cables hopelessly entangled in the process. He needs to have worked out the precise cutting points from one camera to another and to have thought out in advance the type of shots he wants.

All this is worked out during the rehearsals. With the aid of the studio floor plan and models of the sets the director will establish the moves and positions of the actors and the shots will become fixed in his mind. Most directors note this information on their rehearsal scripts and this will eventually be passed to the PA to type in the form of a camera script.

The camera script is therefore a very important document, representing a total handbook of the director's concept of the programme. Certain details might of course change both before and during the studio – shots will alter, cameras change, actors be re-positioned – that is all part of the creative process. But any director going into a multi-camera studio situation without a prepared camera script is asking for trouble, unless of course he has planned for an 'as directed' sequence for a specific reason – perhaps because the shots cannot be anticipated in advance.

So at some stage in the setting up of a production, usually at the latest possible moment, the PA will be handed a document which might well be dog-eared and tatty with more or less legible handwriting scrawled over it. And a word of warning here. It is vital that the PA sets a deadline to the director for the script. Not something unrealistic from the director's point of view, but an agreed date allowing the director sufficient time to prepare the script and the PA time to type it and get copies duplicated.

The front pages

Although the body of the camera script will be made up of the director's instructions typed on to the rehearsal script, the front pages will consist of information worked out either by the PA or in conjunction with the production manager.

The front page of the camera script should contain the programme number, the title, the author and a list of production and technical staff. It should contain the recording date(s), editing and dubbing dates (if known), together with a studio schedule (see page 47).

The body of the document

Now we turn to the main portion of the camera script where the following general guidelines might be helpful:

Colour

The camera script is normally duplicated or photocopied onto yellow paper which allows the type to be clear in the studio yet remain distinctive. But if

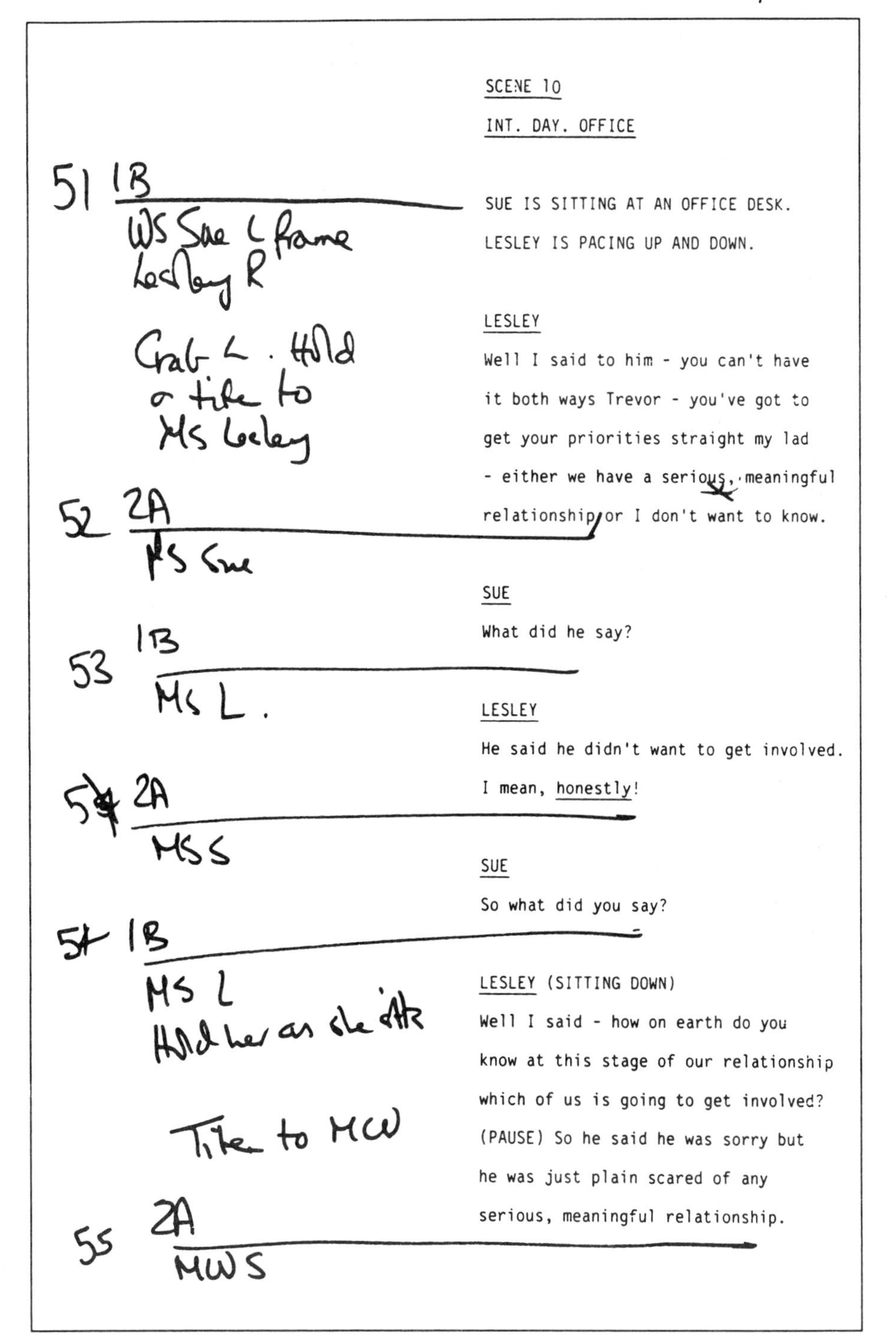

SCENE 10

INT. DAY. OFFICE

51 1B
WS Sue L frame Lesley R

SUE IS SITTING AT AN OFFICE DESK.

LESLEY IS PACING UP AND DOWN.

Crab L. Hold a tilt to MS Lesley

LESLEY

Well I said to him - you can't have it both ways Trevor - you've got to get your priorities straight my lad - either we have a serious, meaningful relationship or I don't want to know.

52 2A
MS Sue

SUE

What did he say?

53 1B
MS L.

LESLEY

He said he didn't want to get involved. I mean, honestly!

54 2A
MSS

SUE

So what did you say?

55 1B
MS L
Hold her as she sits

Tilt to MCU

LESLEY (SITTING DOWN)

Well I said - how on earth do you know at this stage of our relationship which of us is going to get involved? (PAUSE) So he said he was sorry but he was just plain scared of any serious, meaningful relationship.

56 2A
MWS

Example of a draft camera script as it is passed from director to PA.

scenes from a number of different episodes are to be shot on the same studio day then each episode may be identified by being duplicated on different coloured paper.

Numbering

The script is typed out and stapled together in *recording* and *not* in *story* order. This means that fresh consecutive page numbers have to be typed or handwritten on the top right hand corner of each page.

Spacing

Sometimes the PA uses the rehearsal script and just types on the camera details, but more often than not she finds that it is quicker to retype the whole document. It is very important that too much information is not crammed onto one page.

Word processor

It is here, of course, that a software package designed for television proves invaluable as it enables the PA to edit the rehearsal script, add the camera and sound details and alter the layout as necessary.

Consistency of style

Although the layout of the camera script is pretty firmly established, there is plenty of scope for the PA's own, individual style to emerge. Whatever the style, however, it should remain consistent throughout the document.

Boxes

Any technical instructions should be typed inside boxes to make them stand out from the rest of the script. These instructions can refer to cameras, sound, lighting, props, costume, make-up etc. For example:

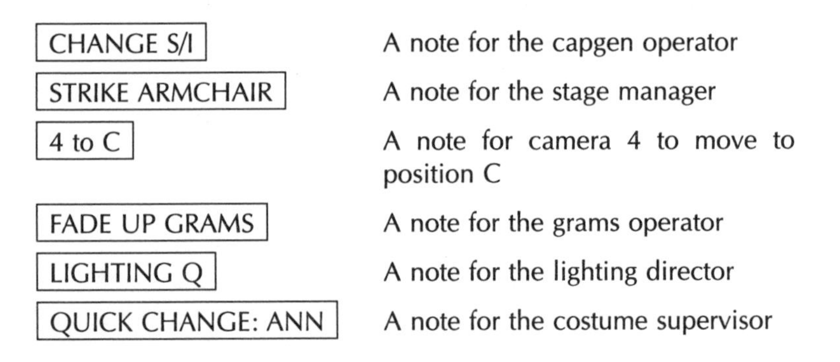

CHANGE S/I	A note for the capgen operator
STRIKE ARMCHAIR	A note for the stage manager
4 to C	A note for camera 4 to move to position C
FADE UP GRAMS	A note for the grams operator
LIGHTING Q	A note for the lighting director
QUICK CHANGE: ANN	A note for the costume supervisor

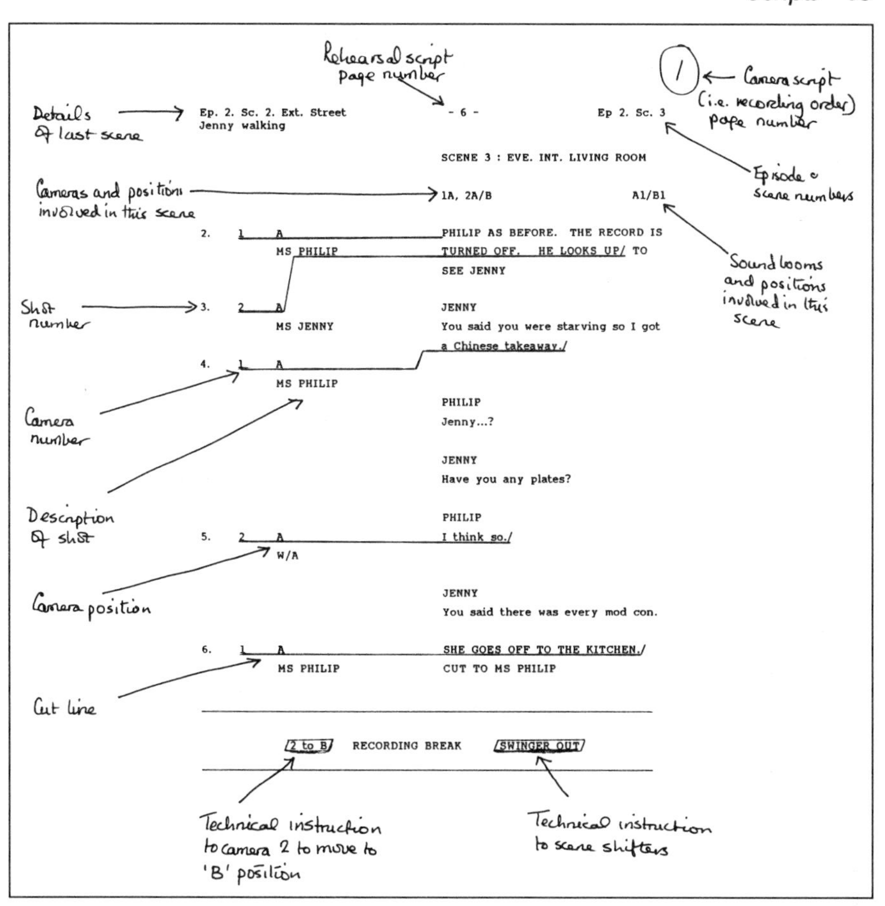

Top and bottom of page

At the top and bottom of each page the PA should give the preceding and succeeding shot number and camera. This will assist her in shot calling, of which more later. For example:

(35 on 4)	At top of page
(43 next)	At bottom of page

Scenes

As with the rehearsal script, each scene should be started on a fresh page, the dialogue and stage directions typed on the right hand side – allowing sufficient space on the extreme right for sound instructions – and the pages should be typed single-sided only.

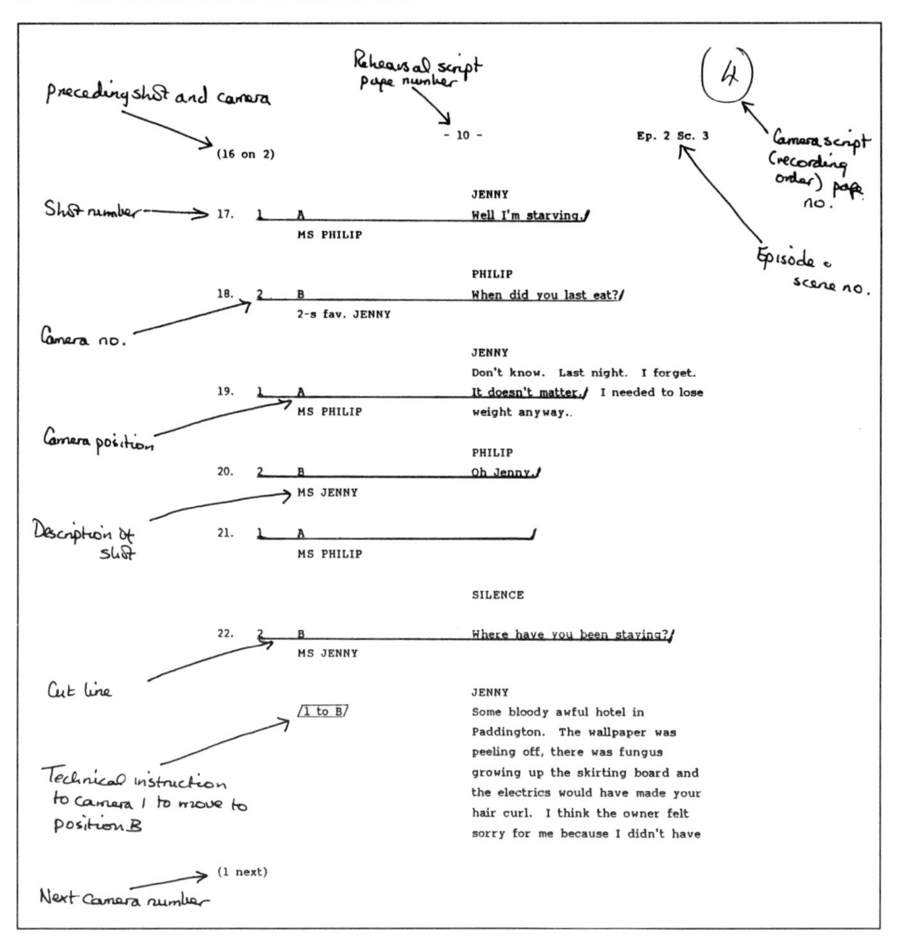

- 10 - Ep. 2 Sc. 3

(16 on 2)

JENNY
17. 1 A Well I'm starving./
MS PHILIP

PHILIP
18. 2 B When did you last eat?/
2-s fav. JENNY

JENNY
Don't know. Last night. I forget.
19. 1 A It doesn't matter./ I needed to lose weight anyway..
MS PHILIP

PHILIP
20. 2 B Oh Jenny./
MS JENNY

21. 1 A /
MS PHILIP

SILENCE

22. 2 B Where have you been staying?/
MS JENNY

JENNY
/1 to B/ Some bloody awful hotel in Paddington. The wallpaper was peeling off, there was fungus growing up the skirting board and the electrics would have made your hair curl. I think the owner felt sorry for me because I didn't have

(1 next)

New scene

At the top lefthand side of the page it is useful to give a brief synopsis of the previous scene as an aide memoire. It is also useful to give a brief synopsis of the scene to come on the bottom righthand side of the final page of a scene.

Shot numbers

The shot number is typed on the extreme left hand side of the page. Every cut is a different shot and should be given a shot number. The numbers should run consecutively in recording order starting from Shot 1. The director might or might not have numbered the shots when drafting the camera script. If they are numbered, they should be checked by the PA when typing the script as it is all too easy to miss one out. If they are not numbered then the PA should do so.

If the camera script is given to the PA in sections, a few scenes at a time as rehearsals progress, she should leave out the shot numbers until she has the complete script.

Camera numbers

Cameras are numbered from one to however many are employed on the production. The director will write the camera chosen for each shot. When the PA is typing the script she should check that the director has not written down two consecutive shots using the same camera or named a camera which is not involved in that scene. She can check this by having a copy of the floor plan to hand.

At every stage of typing the camera script the PA should check for mistakes. This does not mean that she feels her director is in any way a fool but it is terribly easy to make a slip when writing out a camera script, especially when conducting rehearsals at the same time. If the script does contain any irregularities the PA *must* point them out to the director.

Camera positions

After the number of the camera comes the position it is allocated on the studio floor. This is referred to by letter and corresponds with similar letters on the studio floor plan.

If camera 1 starts in position A, then moves to position B and ends up at position C it should be typed as 1 A, 1 B, and so on. The PA must check that the director has allowed time for the camera to change position and as a reminder she should type the move as a box on the script, i.e. | 1 to B. |

Cut line

Directly beneath the camera and its position comes the cut line. This should be typed right across the page and end with a slash upwards at precisely the point at which the director wishes the vision mixer to cut from one shot to another. It should always come at the *end* of the preceding word of dialogue rather than just *before* the new line.

Shot descriptions

Having given each shot a number, a camera and a camera position, the director will then write down a description of the shot he requires. This information should be typed directly under the cut line, in capital letters and inset slightly in order that the camera number is kept clear for the vision mixer to read easily.

A list of commonly used shot descriptions and abbreviations is given on page 76.

```
CORRECT

                              MAN
1.  3  A ____________________ That's it then./

                              WOMAN
2.  1  C ____________________ I told you how it would be./

                              MAN
                              Yes.

INCORRECT

                              MAN
                              That's it then.

1.  3  A __________________ / WOMAN
                              I told you how it would be.

2.  1  C __________________ / MAN
                              Yes.
```

Example of correct and incorrect cut lines. Even if the director has drafted his camera script with the cut line before the new line rather than at the end of the old, the PA should type out the camera script correctly.

If the shot develops, then the further camera instructions should be typed as near as possible to the relevant line of dialogue or stage directions.

Recording break

A recording break signifies that the recording will be stopped. This could be for any one of a number of reasons: to allow a camera to move position; to shift an article of furniture; to move a wall of a set; for a costume change, etc.

In a rehearse/record situation where a scene or number of consecutive scenes are rehearsed and then recorded before going on to rehearse the next block (as opposed to a rehearsal of the whole of the programme followed by a recording), the recording break might occur to allow the next scene or scenes to be rehearsed through.

(50 on 4) - 25 -

	SCENE 10
	INT. DAY. OFFICE
	CAMS 1B/C 2A/B B1 A2
51 1 B	SUE IS SITTING AT AN OFFICE DESK.
WS. SUE L FRAME LESLEY R	LESLEY IS PACING UP AND DOWN.
	LESLEY
CRAB L. HOLD & TIGHTEN TO MS LESLEY	Well I said to him - you can't have
	it both ways Trevor - you've got to
	get your priorities straight my lad
	- either we have a serious, meaningful
52 2 A	relationship/or I don't want to know.
MS SUE	
	SUE
53 1 B	What did he say?/
MS LESLEY	
	LESLEY
	He said he didn't want to get involved.
54 2 A	I mean, honestly! /
MS SUE	
	SUE
55 1 B	So what did you say?/
MS LESLEY	
HOLD HER AS SHE SITS	LESLEY (SITTING DOWN)
	Well I said - how on earth do you
	know at this stage of our relationship
	which of us is going to get involved?
TIGHTEN TO MCU	(PAUSE) So he said he was sorry but
	he was just plain scared of any
	serious, meaningful relationship.

(2 next)

Example of positioning of shot descriptions.

If a break is indicated it should be marked on the camera script by drawing two lines straight across the page. Any technical instructions should be typed in the centre.
Example:

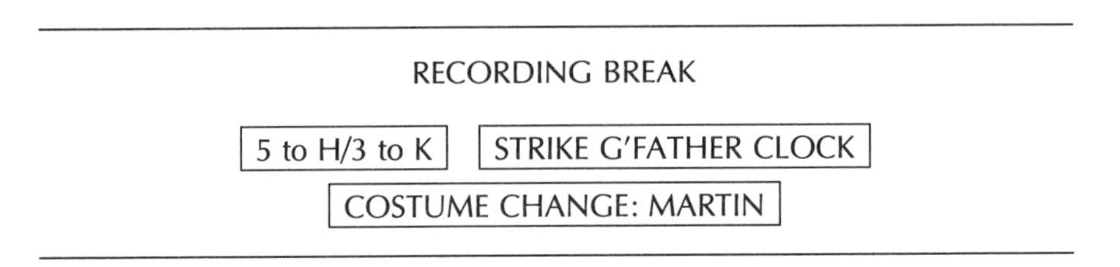
RECORDING BREAK

5 to H/3 to K | STRIKE G'FATHER CLOCK

COSTUME CHANGE: MARTIN

Film or VTR inserts

Film or VTR inserts should likewise be isolated from the main body of the script by means of drawing two lines across the page. They will then stand out more clearly.

'As directed' sequences

If part of the programme is fully scripted and part 'as directed', lines should be drawn across the page for the 'as directed' sequences. Type in the cameras allocated at the top and leave plenty of space for shots and notes to be inserted during rehearsal.

Cameras involved in scene

Type the cameras involved in the scene and their positions directly under the scene heading.

Booms involved in scene

List the booms (microphones extended on long poles) which are involved in the scene on the extreme right hand side of the page, directly under the scene heading. Each boom is identified by letter and its position on the set by number, thus avoiding confusion with the cameras. Therefore the booms involved in a set might be A1, B2 and so on.

Dialogue

The dialogue and stage directions of the scene are as laid out for the rehearsal script, but intersected by cut lines. Don't be afraid of spreading the script so that everything is comfortably spaced. A cramped script will be difficult to follow in the studio.

```
(on 35 on 4)                         /WALL B OUT/

                                     2A/3C/4E                       B2/C3
                                     SCENE 45 : INT. ROOM. NIGHT

36  2A                               JOHN IS ALONE. IT IS VERY      GRAMS
    W/A ROOM                         DARK.  SUDDENLY HIS FACE IS    CAR F/X
    SLOW TRACK ROUND                 LIT  BY THE LIGHTS OF A
    TO MS JOHN                       CAR PASSING IN THE STREET
    /LIGHTING Q/                     BELOW/
37  3C                               JOHN HALF RISES/
    CU JOHN. HE HALF RISES
38  4E                               DET. INSP. (IN DOORWAY)  The
    MS DET. INSP. IN DOORWAY         place is surrounded John.
39  3C                               You haven't a chance./
    CU JOHN
    MIX                              JOHN   All right.  You win.
40  4 E                              You always do, don't you./
    MS DET. INSP.

      /FLOATER IN/          Recording break        2 to B   3 to D
```

Example of a camera script that is far too cramped, incorrectly typed and contains inconsistencies of style.

Sound information

On the extreme right hand side of the page comes any sound information, for example:

- Spot effects (f/x), performed by the assistant stage manager in the studio on cue from the director, i.e. door slams, footsteps.
- Effects on disc, tape, CD or DAT, i.e. bells ringing, clock ticking, hoofbeats.
- Background atmosphere (atmos.), i.e. birdsong, traffic and so on.
- Music played into the studio from disc, tape, CD or DAT.

Much of the sound nowadays is built up in the post production stage during the editing and the sound dub, but there might be certain effects which need to be recorded at the time.

If the sound is to continue for a specific time, a line should be drawn down the side of the page to denote the duration.

```
(on 35 on 4)

      /WALL B OUT/                     SCENE 45 : INT. ROOM. NIGHT

                                       2A/3C/4E                          B2/C3

36.   2  A                             JOHN IS ALONE.  IT IS VERY
         W/A ROOM                      DARK.
         SLOW TRACK ROUND
         TO MS JOHN
                                       SUDDENLY HIS FACE IS LIT BY THE   /GRAMS
      /LIGHTING Q/                     LIGHTS OF A CAR PASSING IN THE    /CAR F/X/
37.   3  C                             STREET BELOW./
         CU JOHN.  HE HALF
         RISES
                                       JOHN HALF RISES/
38.   4  E                           /
         MS DET. INSP. IN DOORWAY      DET. INSP. (IN DOORWAY)
                                       The place is surrounded John.
39.   3  C                             You haven't a chance./
         CU JOHN
         MIX                           JOHN
                                       All right.  You win.  You
40.   4  E                             always do, don't you./
         MS DET. INSP.

      /FLOATER IN/     R E C O R D I N G  B R E A K     /2 TO B/3 TO D/
```

The same piece, properly typed and correctly spaced.

Music

Vocal

You should type out the words of the song in the form of a standard camera script, noting any purely instrumental parts as necessary. If you know how many bars there are to each shot you should type those in, if not you should write them down as your first job in the gallery.

Instrumental only

It is helpful if you can read music and have the score in front of you. You could mark up the score with the shots. If you cannot read music or if there are no

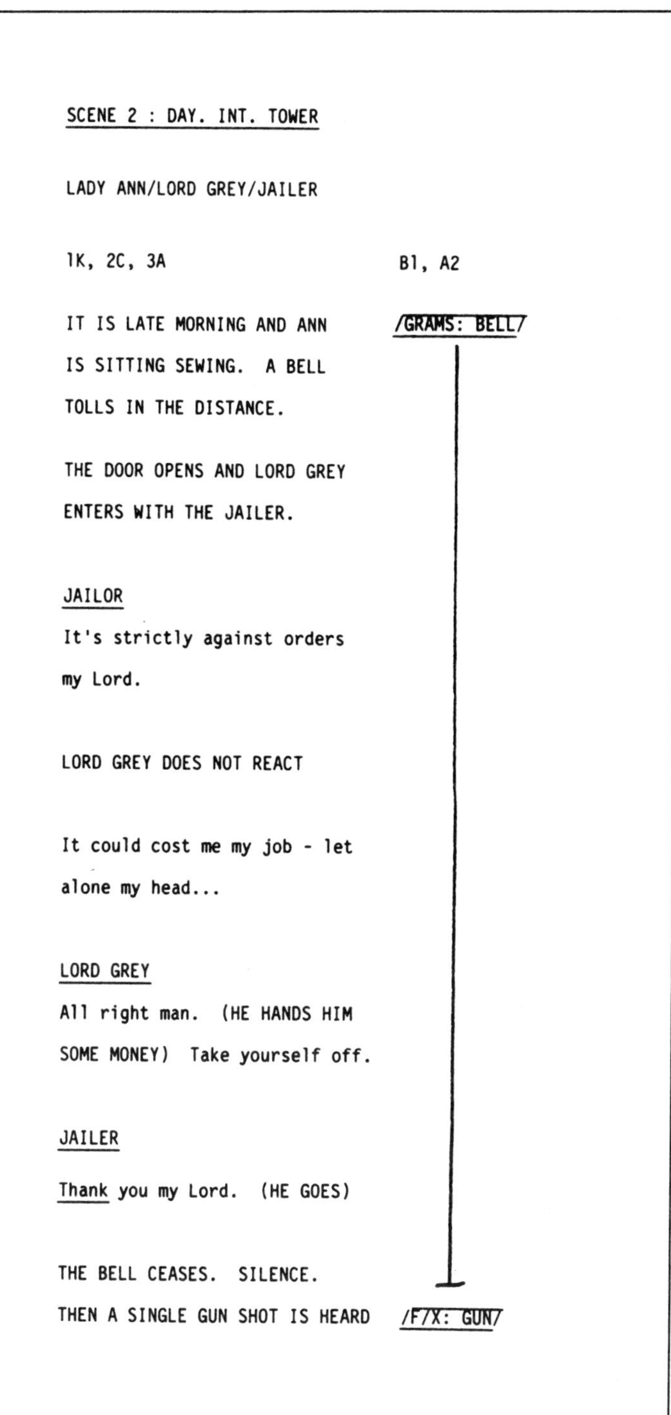

SCENE 2 : DAY. INT. TOWER

LADY ANN/LORD GREY/JAILER

1K, 2C, 3A — B1, A2

IT IS LATE MORNING AND ANN IS SITTING SEWING. A BELL TOLLS IN THE DISTANCE. — /GRAMS: BELL/

THE DOOR OPENS AND LORD GREY ENTERS WITH THE JAILER.

JAILOR

It's strictly against orders my Lord.

LORD GREY DOES NOT REACT

It could cost me my job - let alone my head...

LORD GREY

All right man. (HE HANDS HIM SOME MONEY) Take yourself off.

JAILER

Thank you my Lord. (HE GOES)

THE BELL CEASES. SILENCE.

THEN A SINGLE GUN SHOT IS HEARD — /F/X: GUN/

Example of sound information.

scores available you will have to rely on a form of script which might look something like this:

21. 2 ________________ BASS (12 bars)

L/A BASS PLAYER

CRAB R & WIDEN

22. 1 ________________

W/A. TRACK IN

TO TIGHT 2-s VIOLINS (24 bars)

23. 3 ________________

MS CONDUCTOR

(8 bars)

24. 4 ________________

H/A W/A ORCHESTRA

ORCHESTRA (16 bars)

News scripts

The script on this type of programme is different from most other kinds of scripts in that it forms the *basis* of a working document: it is not a finished document in itself and it is not a complete record of the programme.

You would not, in fifty years time, be able to pick up a news script of today and, solely by reading it, re-live the programme in its entirety. All you would be able to do is read the links between items and gain a sketchy idea about the intervening stories with, perhaps, their overall duration and 'out' words.

The real meat of the programme, the pre-recorded items themselves, are not written in the script. In the case of the news, this is because of the immediacy of news stories. The items would only just have been edited for transmission and there would certainly have been no time and no occasion for a transcript to be made. And of course any elements such as live interviews could not be scripted in advance.

So the script is a mixture of accurately scripted links, exactly as the presenter will read them – unless the presenter is given to unscripted ad libs – scripted voice-over (V/O) commentary to be read over specific news stories and as much information about the pre-recorded items as exist at the time the script is typed.

Stand-by warnings might be typed in as reminders and possibly cues to roll in the various pre-recorded items from their different sources. But many PAs prefer to mark up their own scripts with these cues.

When is it typed?

The links might well be typed by the reporters themselves on a computerized system such as Basys. Printed copies would be made when the links have been checked by the producer and presenter. Otherwise the links would be typed by the PA or a typist as and when they are drafted.

The remainder of the information for the script would be added by the PA. At all events the script will not be typed in chronological order but as and when pre-recorded inserts have been edited. In a news programme, it might well be that parts of the script are still being finalized when the programme has gone on air. On many programmes of this nature, the finished pages are never finally stapled together.

Page numbers

Because the script is typed in this way, consecutive page numbering becomes meaningless and each section is referred to, both in and out of the studio, by its item or sequence number, each sequence comprising the individual news story plus its link.

Each sequence must therefore start on a fresh page and the only numbering that usually occurs is when more than one page relates to the same sequence.

Layout

The standard layout for camera scripts is one where the picture details are placed on the lefthand side of the page and the dialogue on the right. The pages are typed on one side only and camera scripts are duplicated or photocopied on coloured paper.

It is extremely important to allow plenty of space when typing the script as the vision mixer, director and PA will all be marking up their copies and plenty of room is appreciated.

Lefthand side

The visuals are usually typed on the lefthand side of the page, in capital letters and the line across the page denotes a cut by the vision mixer to another shot whether it be to another camera's output or to another source, i.e. VTR etc.

Above the line the camera or source is marked and below the line an indication is given of the picture.

Standby warnings and instructions are also marked on the lefthand side of the page, usually framed in boxes so that they stand out.

The duration of the pre-recorded item is typed in, if this is known, together with the 'out' words.

Righthand side

The presenter's links are typed out in full in double spacing on the righthand side of the page. These are identical to the words seen by the presenter in the studio and read by him or her from the teleprompt. If any changes are made, the PA must note them on her own script as they might affect the cueing in of different sources.

	SEQUENCE 9 : HIDDEN TREASURE
/S/B VT/	
CAM 2 IN VISION	ANDREW Hidden treasure. The very thought of it conjures up visions of gold coins, silver tankards, the rich and precious heritage of days long past. But hidden treasure can be of a very different kind as Mrs Mary Watkins found when she sent off a parcel of old clothes to the Red Cross in response to the latest aid appeal. Chris Johns reports:
VTR : HIDDEN TREASURE CAPGEN : Mary Watkins at	Dur: Out words : ....
CAMS 2 & 3	/Sport link next/

OK Seq: Storyline 52 Source Jrn Dur: Total

AXP 52*BATTLE OF POWICK INTRO DAVID 0:16

Fri Sep 18 16:53 houghton +

CAM CUE DAVID IN VIS______________/It's been billed as the biggest cavalry charge on English soil since the seventeenth century. Ten thousand are expected to turn out to watch. A thousand will take part. What's it all about? John Yates takes a three hundred and fifty year trip back in time in Worcester.

BETA FOLLOWS>>>>>>>>>>>>>>>>>>>>>>>>>>>>

OK Seq: Storyline 53 Source Jrn Dur: Total

XXP 53*BATTLE OF POWICK/YATES BETA+SUPERS E1:25

Fri Sep 18 16:52 houghton +

BETA______________________/

IN :

DUR:

OUT: ....A DUMMY RUN."

SUPER @

RICHARD TOWNSLEY

BATTLE ORGANISER

SUPER @

NATHANIEL FIENNES

PRINCE RUPERT

If the examples above are studied it will be clear that this type of script does not contain a great deal of information. Apart from the links, the information given is patchy and the little contained is, needless to say, subject to change and alteration.

Providing the PA is working on a programme where she is expected to give standby warnings and roll in pre-recordings, as soon as she receives any part of the script she begins marking it up.

If the inserts contain a lot of name supers it might be better to place the insert details on a fresh sheet, rather than trying to cram it all onto the same page as the link.

Computerized system

Scripts which are typed using a computerized system will probably be laid out somewhat differently to the above. The examples given above were typed using the Basys system and you will notice that the link and the insert appear on different pages.

Shot descriptions

The following gives a list of the most commonly used shot descriptions and abbreviations:

W/A	Wide angle shot. Such a shot takes place in a wide area of the set in front of the camera. It is sometimes referred to as a VLS (very long shot).
LS	Long shot. A shot which directs the viewer's eye to the depth rather than the width of the shot.
MLS	Medium long shot. Refers to a shot comprising the head to just below the knee of the subject.
3-s	Three shot. A shot containing three central characters.
2-s	Two shot. A shot containing two central characters.
2-s fav. X	A shot with two people – the camera favours the person facing the camera.
o/s 2-s	Over the shoulder two shot. Two people are seen in the shot but the camera is looking at one of them over the shoulder of the other.
Mid 2-s	Comprising the head to just below the waist of two people.
Close 2-s	Comprising the head and shoulders of two people.
MS	Mid shot. A scene at normal viewing distance. In the case of a human subject the camera frame cuts the figure just below the waist.

MCU	Medium close up. The camera frame cuts the figure at chest level.
CU	Close up. The camera frame cuts the subject just below the neck.
BCU	Big close up. The face fills the frame.
X's POV	X's point of view shot. The camera is X and sees as if from his point of view.
H/A	High angle. The camera is above the action and looking down on it.
L/A	Low angle. The camera is below the action and looking up.

Camera movements

PANNING	Camera turns from one side to the other, pivoting horizontally on an axis, either right to left or left to right.
TILTING	Camera pivoting vertically on an axis, tilting up or down.
TRACKING	Camera is physically moved forward or back, towards or away from the subject.
CRABBING	The camera is physically moved crab-wise or sideways to the direction of view.
CRANING AND JIBBING	A movement by a camera mounted on a crane dolly. The dolly has a jib arm which can be raised and lowered rotating around its fulcrum.
Z/I	Zoom in. The camera is not moved but the focal length of the zoom lens is increased. This magnifies the subject without changing the perspective of the scene (as opposed to a track where the camera moves towards the subject and the perspective changes as if you were walking towards it.)
Z/O	Zoom out. The lens is adjusted in the reverse direction from the above.

Abbreviations relating to action

A/B	As before
FAV.	Favouring
F/G	Foreground
B/G	Background
F/WD	Forward
B/WD	Backward
X/s	Crosses or across
CAM R	Camera right, i.e. as seen from the camera's – and the viewer's – position when facing the action
CAM L	Camera left
O/S	Over the shoulder
OOV	Out of vision

OOF (L or R)	Out of frame (left or right)
Q	Cue
S/B	Stand by

Abbreviations relating to sound

F/U	Fade up
F/D	Fade down
MUTE	Without sound
V/O	Voice over
S.O.F.	Sound on film
S.O.T.	Sound on tape
S.O.VT	Sound on videotape
S & V	Sound and vision
MIC	Microphone
GRAMS	Music or sound effects from gramophone records. It is also used as an instruction to the person known as the 'grams operator', therefore could also refer to music or sound effects on tape, compact disc or DAT.
TAPE	Music or sound effects solely from tape recording
F/X	Effects
SPOT F/X	Sound effects made in the studio
ATMOS.	Atmosphere
UP SOUND	Bring up sound effects

Instructions for vision mixer

F/I	Fade in
F/O	Fade out
S/I	Superimpose
T/O	Take out
WIPE	
MIX	
CUT	

Abbreviations relating to graphics or video effects

DFS	Digital frame store
DVE	Digital video effects
CK	[Chromakey. Replacing part of an electronic picture with material from another source. In some companies this is known as CSO (colour separation overlay).
CPU	Caption projection unit
S/F	Slide file.
GFX	Graphics effects

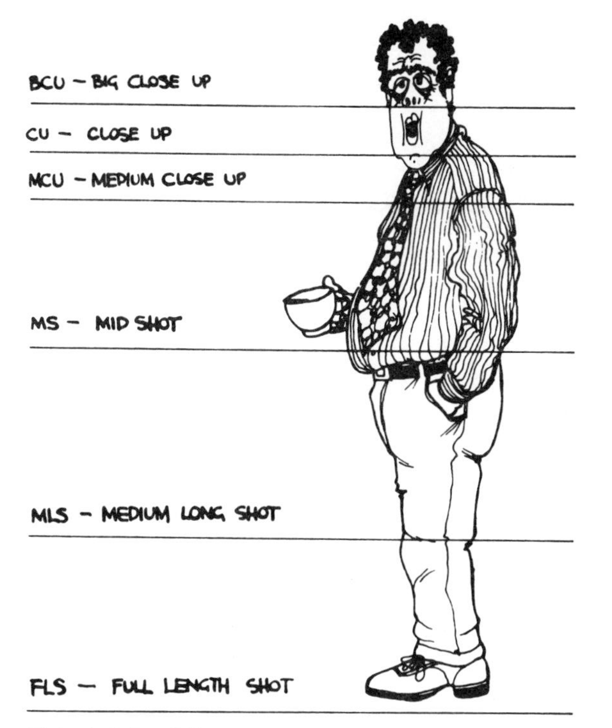

Main shot descriptions.

Camera cards

Once the camera script has been typed and checked it should be photocopied or duplicated and made available to everyone on the production who needs it. The cast will not need copies of the camera script.

The PA prepares camera cards for the camera operators. These are ideally typed on thin card or on paper which is stapled to cards. The typing should be clear enough for reading under possibly dim lighting conditions.

Each card contains only the information relevant to that particular camera: camera 1 would receive cards with the minimum details of camera 1's shots and so on. The shots are extracted by the PA from the main body of the camera script.

Type on the card the shot numbers relevant to that camera, the camera's position and a description of the shot. Do keep the description brief and unambiguous no matter how your director has phrased it on the camera script. For example:

On camera script:

START TIGHT ON HARRY, FRAMING HIS HEAD AND SHOULDERS. WIDEN SHOT AS HE WALKS R TO DRINKS CABINET AND PAN WITH HIM.

Should read on the camera card:

MCU HARRY WIDEN & PAM HIM R (TO DRINKS CABINET)

Keep it brief and concise.

Every episode, scene and set should be marked on the camera card as well as the numbers of the other cameras operating during that scene. Recording

CAMERA : One   Page 3   TITLE : JOE AND HARRY

Shot	Pos.	Description	Notes
		SCENE 2 : DAY. INT. HALL (CAMS 1A/B, 2C, 3A)	
9	A	2-s JOE/MARY TRACK WITH JOE AS HE X'S L. LOSE MARY Tighten [illegible] WITH MOVE CRAB LEFT TO MISS SETTEE	
14	A	MS JOE (AS HE TURNS)	
14B		A/B (HE TURNS AWAY) /T TO B/	
24	B	DEEP 3-s FAV. JOE PAN JOE R & TIGHTEN INTO 2-s. JOE SITS.	(WATCH IT HE SITS FAST!)

RECORDING BREAK

Example of a camera card.

breaks, tape run-ons and VT inserts should also be included on the cards of all the cameras involved.

The typing *must* be well spaced out. Don't type too much on a card and don't put a fast shot change at the bottom of a card. It is far better to leave half a card blank if necessary.

Do leave plenty of room on the righthand side of the card as well as underneath each shot for notes to be added. This is especially important when 'as directed' sequences are involved.

Distribution and use

The PA should give the sets of cards to the camera operators. Cameras with more than one operator, i.e. cameras on dollies or cranes, will need additional sets of cards.

During the studio the camera operators will be listening to the PA calling the shots (see Part Three chapter three) and will have a clear reference on their cards as to the point that has been reached in the script – by virtue of the shot number called – the number and description of the shot for which they will next be required and the position to which they should move.

It is also useful to remind camera operators more than once of the shot they are on and what is coming next during shots of lengthy duration.

9 Graphics

The increasing use of computers and new technology with specialized electronic effects has, over the last few years, made a tremendous impact in the field of graphic design. The range and scope of advanced computerized effects are staggering and it is easy to become confused and uncertain of what and how different effects can be achieved.

Because this book is primarily for PAs and because it is the PA who will most probably have to book facilities and liaise with the graphic design department, I will briefly work through the graphic design element in any programme, whether live or recorded.

Lettering (end credits and name superimpositions)

Every single production will need lettering of some sort, even if it is only the end credits. Lettering is the basic bread and butter of graphic design.

Lettering is achieved by means of computers known as character generators. These provide a comprehensive video lettering service capable of producing letters in a variety of type-faces, size, shape and colouring. The lettering can either be assembled and edited in advance of the studio and stored on disc, or an operator can type up the letters in the gallery of the studio for recording or transmission if live.

In addition to lettering, computers will produce logos and symbols. A 'house' style can be worked out for each programme with an overall design of logos, symbols and type-faces. The design for each programme can be stored on floppy disc.

Most television companies refer to captions created from character generators by the trade name of the system chosen by each company i.e. Aston, Abekas, etc. For the purpose of this book I shall call them 'capgens'.

The PA will need to provide the graphic designer with details of the front and end credits and any name superimpositions. It is naturally very important for the PA to check the spelling of the actors', production members' and anyone whose name appears on screen as she will not be at all popular if names are misspelt. She should also check abbreviations: someone known as Chris might well wish his name to be Christopher on the credits.

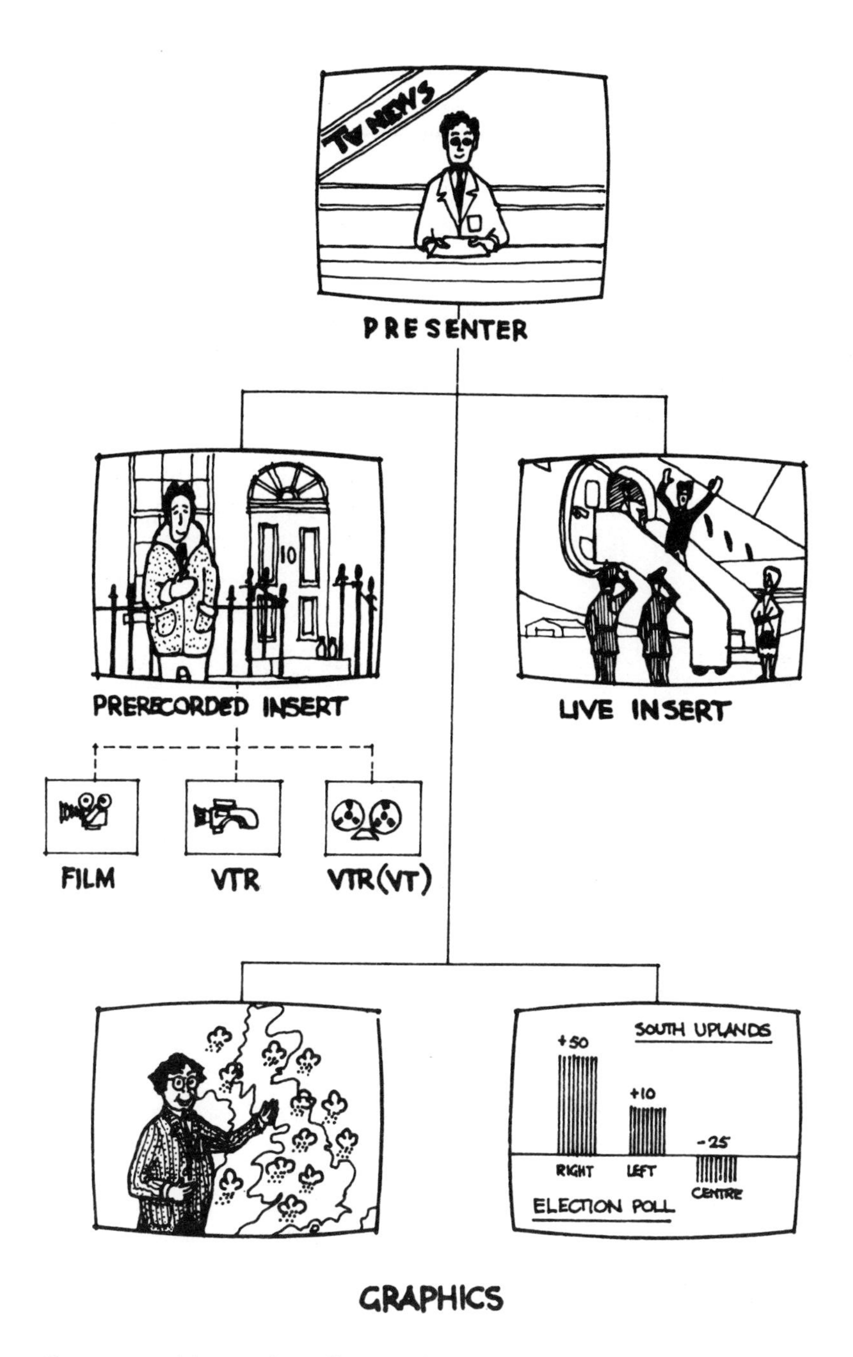

Components of the news/topical/current affairs programme.

Captions (maps, diagrams, photographs recorded in the studio)

Many productions require maps, diagrams or still photographs. These could be recorded using a rostrum mounted video camera.

Specialized computers, can store, re-touch, re-size and re-position captions as well as present them in the required order. As the scope of these computers embrace more than just maps, diagrams and still photographs, we will look at them in conjunction with the next element in graphic design.

Creative sequences (titles, animations, inserts, promotional trailers)

This is the main-stream work of the design department and each job usually demands innovation and originality. Each job is therefore by definition different. There are various computers on the market which help both in the creation and the storage of the work and which can make hitherto impossible effects both simple and easily achieved.

There are computers designed for creating fine art and graphics, i.e. Paintbox, which are rather like an electronic version of paint, paper, paste, scissors, card and stencil to the graphic artist. There are electronic systems providing some or all of the following: multiple pictures; borders and matte effects; images at any size from normal to virtually zero; wipes; split screens; flips; cubes; re-sizing and re-positioning of still pictures. All these effects can be recorded and stored. Then there are frame stores (S/F – slide file and DFS – digital frame store) which electronically store any picture in whatever order is selected by the operator. There are also digital picture archives (DPA) which act as a storing house for any pictures used on past programmes.

But it must be remembered however that all these computers are specialist machines with fixed software programmes. They are ultimately as good or as bad as the graphic artist who operates them. In order to make the fullest and most creative use of these computers it is essential that the graphic designers be involved at an early stage in the programme.

10 Run-up to 'live' and 'as-live' programmes

Components of programme

Before looking at the run-up to this type of programme, we ought perhaps to look at the different elements that go to make up that programme. These can be roughly itemized as follows:

Presenter(s)/newsreader(s)

No programme of this type would be complete without the presenter (newsreader, linkman or anchorman or whatever he/she is called). There may be one, two or even more on a programme and their role is crucial in introducing the different items, providing smooth links from one item to another, conducting interviews or giving live voice-over commentaries to pre-recorded inserts. In other words, they hold the programme together.

They are highly professional people and as time-conscious as the PA. The consistency of their speed of delivery is vital as inserts are cued in according to the duration of their links and, providing the PA in the gallery has done her work correctly, the presenter bears a great responsibility for the smooth transition between items. They may, and frequently are, called upon while on air, to ad lib for a length of time in order to pad an under-running programme or, conversely, to cut down on prepared links to save time if over-running. A good presenter is invaluable to a PA and a good PA is a life-line to a presenter.

So, if the presenter is the first component in this type of programme, the next must be:

Pre-recorded inserts

In the past the inserts in a news magazine programme would have been almost entirely shot on film, normally reversal, the film being processed and edited immediately before transmission. With the advent of portable single camera systems where a video camera was either plugged directly into a portable video recorder, e.g. High Band U-Matic, or, later, when the VTR was built into the camera itself, e.g. Betacam, the film insert gave way to the videotape insert.

Because the personnel involved in making these inserts originally came from a film background, a number of terms sprang up to differentiate these portable video systems from mainstream (2″ or 1″) videotape originating from studios or outside broadcasts and staffed by personnel with an engineering background. With the advent of the high quality full bandwidth pictures available from the Betacam SP format, common sense has prevailed in the industry and all inserts are generally known as VT or VTR. In the few cases where an insert has been shot and edited on film it will almost certainly have been transferred to videotape prior to the live transmission.

For the purposes of this book I have referred to all pre-recorded inserts as VTR but some companies may refer to the insert by the format rather than the term VTR, e.g. Beta, 1″, D2 etc. You may occasionally come across terms such as PSC, ENG and EFP but these are increasingly going out of use.

Stories for the pre-recorded inserts are gathered in various ways, by researchers, by agencies, by 'our man on the spot', by events that have been ear-marked long before by the producer, or by a number of other means.

If the programme is live or 'as-live', these inserts will be injected directly into the studio at the appropriate points, or they may be edited together with the studio recordings at a later stage.

Live

There might be live injects coming from other studio centres, outside broadcast units or satellite. These live injections are very difficult to schedule accurately in the overall running order of a programme, very often because a news story is just about to break or has just broken.

A classic example of this is of a news bulletin extended to give live coverage of the arrival of a plane containing some important person. A cold reporter, shivering on the tarmac of the airfield ad libs his way through ten minutes of agonized waiting for the plane which has been delayed. The cameraman shows us shots of everything he can find in his viewfinder, whether or not it bears any relationship to the subject. The wait continues. At last the director cuts back to the studio and at that precise moment the plane lands. Live programmes are unpredictable if nothing else!

Graphics

A fourth and most important element of any news/topical current affairs type programme is that of graphic design.

Run-up to studio

If we follow through a day in the life of a PA on a fairly typical news-type programme that is to be transmitted early in the evening, we might envisage the following schedule:

DRAFT RUNNING ORDER WEEKDAY NEWS FRIDAY 27 NOVEMBER 1992

In : 18.00.00
R/T : 20.25
Out : 18.20.25

Prog. No: XYZ/1234/A

Producer : MARY JONES
DIRECTOR : DOUGAL McLEAN
P.A. : JOSIE BLANE
F.M. : FRED SMITH

Presenters : CHERRY STONE
ANDREW PLUM

SEQ.	SOURCE	TITLE/PRESENTER/REPORTER	EST. DUR.
1	VTR (A) STUDIO (Cams 1/2)	TITLES Cherry and Andrew	0.23 (0.30)
2	STUDIO (1/3) VTR (B) + Aston	Cherry links to LOST CHILDREN (PT)	(2.35)
3	STUDIO (2/4) VTR (C) + Aston	Andrew links to NEW TECHNOLOGY (TW)	(2.20)
4	STUDIO (1) VTR (A) + Aston	Cherry links to SOCCER VIOLENCE (LW)	(1.15)
5	STUDIO (1/2) VTR (B) (Mute) VTR (C) + Aston VTR (A) (Mute)	Andrew and Cherry SHORTS HOUSING PROJECT LOIS GREY BARN MURDER	(2.45)
6	VTR (B)	TEASE : HIDDEN TREASURE	(0.15)
7	STUDIO (2/3) VTR (C) + S/F's	Andrew links to WAR CRIMES (CP)	(2.10)
8	STUDIO (1/4) VTR (B)	Cherry links to STATE VISIT (LE)	(1.25)
9	STUDIO (2) VTR (A) + Aston	Andrew links to HIDDEN TREASURE (AB)	1.55
10	STUDIO (2/3) + Aston VTR (B)	Andrew links to SPORT	(2.35)
11	STUDIO (1) VTR (C)	Cherry links to WEATHER	(1.15)
12	STUDIO (1/2) + Aston	Cherry and Andrew CLOSING HEADLINES	(0.35)
13	STUDIO (1/2) VTR (A) + Roller	AND FINALLY . . . Cherry GOODNIGHT (C & A) PROGRAMME CLOSE	 (0.35) (0.15)

STANDBYS : RAIN FORESTS (TP) 2.10
INDUSTRIAL REVIVAL (CP) 1.10

Running order

The draft running order is compiled by the producer of the day, (see Part Two chapter six) and distributed by the PA.

It becomes clear that, at this stage of the proceedings, all the timings other than the opening and closing titles and Sequence 9 in our programme are *estimates* only, given by the producer, who arrived at them in the following way:

- Firstly on the basis of the information given at the initial meeting, but it must be remembered that at this stage the inserts have not been edited.
- Secondly by the producer's own feelings about the worth, in terms of screen time, of each item.

It cannot be said too often that any item might be changed, dropped, shortened or lengthened at any time until the programme is off the air. In the event of a major news item breaking, the entire proposed running order might be scrapped.

The following things happen at no specified times during the afternoon:

Pre-recordings

The director makes any necessary pre-recordings in the studio. This may involve the services of the gallery PA or some other PA, who will then pass on the following information to the gallery PA:

- duration of pre-recorded item, and
- the 'in' and 'out' words and picture.

Completion of items

Journalists will be busy editing their own individual stories. Once completed it is, in theory, either their responsibility or the responsibility of the VTR editor, to give the PA the following information for that item:

- Duration of item.
- The 'in' and 'out' words and picture.
- Details of name superimpositions and the timing at which they should occur.
- If known, which channel is to play the item into the programme.
- Any music or other copyright details.

But frequently journalists and VTR editors are hard enough pressed to make the deadline of transmission without worrying about information for the PA. It then falls on the PA to chase up these details.

This is where a computerized system such as Basys is so helpful to the PA, as the information she requires appears directly on the VDU.

Graphics

During the afternoon the graphic element of the programme will be compiled in the correct order for the programme.

Capgens

A list of capgens will be made and given to the operator, again in the correct order. In some instances the PA will operate the caption generator herself.

Links

As items are edited, the presenter's links will be written and passed to the PAs or secretaries for typing, unless a computerized system is used. These links, once they have been agreed by the producer and presenter, form the basis of the script.

Teleprompt

As the script is duplicated, a copy should be passed or downloaded to the teleprompt operator.

Floor plan

The director will work out his camera and boom positions within the studio set and plot them on a floor plan. (See page 55 for illustration of a floor plan.)

The pace hots up

From mid-afternoon onwards the pace increases in the newsroom as the transmission time draws near. As the afternoon progresses the gallery PA will be able to add to her running order or time chart the *real,* as opposed to the *estimated* timings, and keep an eye on the real running time as opposed to the estimated one (see next chapter).

Sometimes the PA also works out which VTR channels will run the various inserts. In doing this she should always bear in mind that it takes a minimum of 35 seconds to eject a cassette from, say a Beta SP machine, and insert another one.

Eventually the final running order is printed out or typed and duplicated – often on different coloured paper from the draft – and distributed.

The PA's job

The gallery PA is very much at the centre of all this activity. She is the one who, in many ways, holds the various threads together. She is the central point of contact, constantly being given information, assimilating it, passing it on to those who need it and, above all, keeping a keen eye on the overall timing of the programme as the timings of the individual items become finalized.

It is useful, but not standard practice, for the gallery PA to sit with the producer and director during the afternoon as the accent is on fast, accurate communication between these three people.

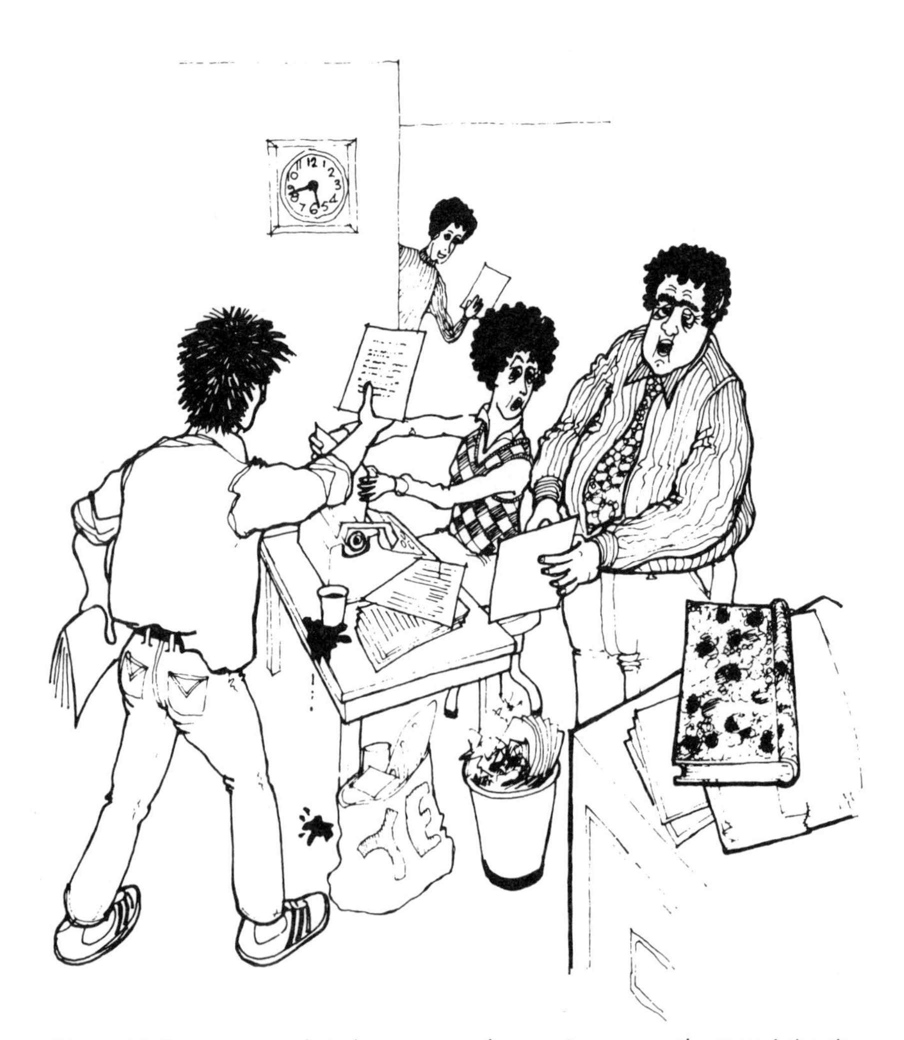

'From mid-afternoon onwards in the newsroom the pace increases as the transmission time draws near.'

Liaison

The key to the gallery PA's job during the run-up to the programme is liaison. Liaison with the producer and director throughout the afternoon is essential both to keep them aware of changes as they are brought to her attention and to keep up to date with the fluctuations of the overall duration.

That is not to say that the PA will run to the producer every time her sums make the programme five seconds or so light or heavy, but in a programme of, say, twenty minutes length, the producer will want to know if it is running around two minutes over or under. Exactly when to bother the producer is very much a matter of experience and of having an understanding of the content of the items.

Communication is a two-way process, however, and it is right for the PA to expect to be kept informed of any changes planned by the producer or director during the run-up to the programme.

Timing

But above all, the PA working on live, fast, news-type programmes *must* be able to add up and subtract time quickly and accurately, especially when it is only one minute to transmission and a flood of completed items with their durations is handed in – and some PAs would say that if the information comes in one minute before transmission it is going to be a relaxed show!

11 Timings

As the real times replace the estimated ones on the running order, the PA can tell whether the programme is over or under-running at any point during the setting up of the programme. You will acquire the real timings in the following way:

Real timings

Pre-recorded items

Using a computerized system, the journalist will enter the duration of the item once it has been edited, together with information concerning capgens. If there is no computerized system, this information might come to the PA by means of a source sheet which is filled in by the editor. Each item has its separate sheet and on it is given the story title, sequence number, channel from which it is to be generated if known, sound details, capgen details, the 'out' words and overall duration.

Links

The links into (and sometimes out of) the pre-recorded items are written by the journalists. Once they have been written, they should be timed. It is standard practice to count three words to a second when timing the links.

The duration of the insert, together with its link, should be added together to give you a 'real' as opposed to an 'estimated' sequence time and the results noted on a time chart.

Time chart

From the example on page 94, you will see a number of columns.

The first is for the sequence number and title. The second is for the producer's estimated timings. This comprises two figures, the duration of the scripted link and the duration of the pre-recorded insert. Next to that comes the cumulative times, making an overall programme duration of 20.25″. When the actual times are given to you (that is the duration of the pre-recorded insert plus the link which you have timed yourself), you enter the figures in the next column. We shall come to the last two columns later.

SEQUENCE: 9

STORY: RAILWAYS

CHANNEL: VTR- 6
(if known)

SOUND (v/o times etc.)	CAPGEN (Time and name)	ANY OTHER DETAILS
S.O.VT	@ 16" 'MICHAEL WRIGHT' @ 42" 'ALAN ROWSE'	HOLD ON LAST SHOT UNTIL TRAIN ENTERS TUNNEL (at 1'35")

DURATION: 1'35"

"OUT" WORDS: "... there is no going back" (See above)

Example of a source sheet.

TIME CHART

TITLE	EST. TIME	CUM.	ACTUAL	BACK	ON AIR
1. TITLES	0.23 / .07 } .30	.30			
2. LOST CHILDREN	20 / 2.15 } 2.35	3.05			
3. NEW TECHNOLOGY	20 / 2.00 } 2.20	5.25			
4. SOCCER VIOLENCE	15 / 1.00 } 1.15	6.40			
5. SHORTS	20 / 2.25 } 2.45	9.25			
6. TEASE	0.15	9.40			
7. WAR CRIMES	20 / 1.50 } 2.10	11.50			
8. STATE VISIT	25 / 1.00 } 1.25	13.15			
9. HIDDEN TREASURE	15 / 1.40 } 1.55	15.10			
10. SPORT	15 / 2.20 } 2.35	17.45			
11. WEATHER	1.15	19.00			
12. CLOSING	.35	19.35			
13. GOODNIGHT/VTR	.50	20.25			

Example of time chart.

Working on the time chart, let's say that the first accurate duration given is for:

SEQUENCE 10. The accurate timing is 2.50″. The estimated timing is 2.35″. Take 2.35″ away from 2.50″ and we are left with 15″. This makes the programme, at this early stage, over-running by 15″. Note this at the bottom of the page.

TIME CHART

TITLE	EST. TIME	CUM.	ACTUAL	BACK	ON AIR
1. TITLES	0.23 .07 } .30	.30			
2. LOST CHILDREN	20 2.15 } 2.35	3.05	2.05		
3. NEW TECHNOLOGY	20 2.00 } 2.20	5.25	?		
4. SOCCER VIOLENCE	15 1.00 } 1.15	6.40	1.45		
5. SHORTS	20 2.25 } 2.45	9.25			
6. TEASE	0.15	9.40	.15		
7. WAR CRIMES	20 1.50 } 2.10	11.50			
8. STATE VISIT	25 1.00 } 1.25	13.15	2.05		
9. HIDDEN TREASURE	15 1.40 } 1.55	15.10	1.55		
10. SPORT	15 2.20 } 2.35	17.45	2.50		
11. WEATHER	1.15	19.00	1.20		
12. CLOSING	.35	19.35	.35		
13. GOODNIGHT/VTR	.50	20.25	.50		

Example of time chart.

We then receive the timing for:

SEQUENCE 6. This is 15″, the same as the estimate.

SEQUENCE 11 is 1.20″. This is 5″ over the estimated time. Add this to the 15″ over-run we already have and now the programme has spread by 20″.

SEQUENCE 4 is 1.45. This is 30″ over the estimated timing and brings the overall programme spread up to 50″.

SEQUENCE 3. We hear that this will be late. It could become what is known as a 'floater', that is, fitted into the programme whenever it is completed. Its estimated duration must therefore be kept in mind if it is not ready when the programme goes on air.

SEQUENCE 8 is completed. Its duration is 2.05″, which is 40″ over the estimated time. The overall spread is now 50″ + 40″ = 1.30″.

You are about to inform the producer when the timing comes in for SEQUENCE 2. This is 2.05″, instead of the estimated 2.35. The overall spread is now back to 1.00″.

SEQUENCES 9, 12 and 13 are as estimated.

At this time we know that the overall spread of the programme is 1.00″. This might not matter if the programme is to be recorded 'as live' and edited to length afterwards. But if it is important that the programme fits into a specific time-slot as it is being transmitted live, then such programmes always have built-in ways for the PA to adjust the timings in advance or while on air.

Fixed items and buffers

Because we break a programme down into its component parts, its different sequences, we find that these fall into two distinct groups:

- items of fixed duration,
- items of variable duration (buffers).

The items of fixed duration are those that have been pre-recorded and edited. It might be possible to come out of one of these items early, but generally speaking the stories will either have to run their course while on air or be dropped from the programme altogether.

It might also be possible to shorten or lengthen the links into these items but these are fixed to a certain extent, especially if the PA needs a certain length of time in which to pre-roll the inserts. (See p. 101.) There is also the risk of confusing the presenter if the links are changed arbitrarily while on air.

Items of variable duration, known as 'buffers', give the producer leeway in adjusting the overall timing without having to resort to such drastic measures as dropping an entire item.

The buffer is always a 'live' item. It could be an interview, it could be the closing headlines, it could – and often is – the weather, it could be a cake-

TITLE	EST. TIME	CUMULATIVE
1. TITLES (VTR)	0.10	
2. AMERICA REPORT (VTR)	1.30	1.40
3. REPORT FROM GREECE (VTR)	2.00	3.40
4. " AUSTRALIA (VTR)	.40	4.20
5. INTERVIEW	3.00?	7.20
6. AROUND THE WORLD (VTR)	1.10	8.30
7. STUDIO DISCUSSION	6.00?	14.30
8. CLOSING (VTR)	0.40	15.10

Example of timing. In this example, the items of variable duration, the buffers, are 5 and 7 (the interview and studio discussion). Everything else is pre-recorded and of fixed duration. When the real times for the pre-recorded items are known, the buffer durations can be altered to fit in with the overall running time of the programme. The essential time, in the above example, would be 14.30", the end of the second buffer. It would be vital to hit that time correctly so that the closing credits, pre-recorded on VTR, could be shown in their entirety. If the programme has only a limited number of play-in machines there may be technical problems in changing the item at short notice if sufficient time is not allowed to reload the machine with the required item.

making demonstrating in a magazine-type programme. Whatever it is, its duration can be changed while on air, provided that the producer has been informed by the PA by how much the buffer is required to expand or contract.

Let us look at an example. A producer has estimated 3.00″ on a live interview, but just before we come to the interview the PA realizes that the programme is over-running by 50″. The producer might then allot 2.10″ to the interview, thus checking the over-run and bringing the programme back to time.

The buffer item therefore is a means of getting the programme back to time if it is over- or under-running. However it often happens that the buffer itself will over- or under-run. The interview which was allotted only 2.10″ spreads to 2.24″, thus making our programme still over-running by 35″. So long as the producer is told, he or she can then decide whether to save time in another buffer item, drop a fixed item altogether or ask the PA to throw herself on the mercy of presentation to allow the over-run.

Once we have a few accurate timings on our time chart, we are able to start back or forward timing.

Back and forward timing

Backtiming

Backtiming enables you to go on the air with the minimum of information. Because the aim is to get off the air smoothly and on time, in one sense the timings at the front of the programme are relatively less important than those at

TIME CHART

TITLE	EST. TIME	CUM.	ACTUAL	BACK	ON AIR
1. TITLES	0.23 .07 } .30	.30			
2. LOST CHILDREN	20 2.15 } 2.35	3.05	2.05		
3. NEW TECHNOLOGY	20 2.00 } 2.20	5.25	?		
4. SOCCER VIOLENCE	15 1.00 } 1.15	6.40	1.45		
5. SHORTS	20 2.25 } 2.45	9.25			
6. TEASE	0.15	9.40	.15		
7. WAR CRIMES	20 1.50 } 2.10	11.50		10.50	
8. STATE VISIT	25 1.00 } 1.25	13.15	2.05	12.55	
9. HIDDEN TREASURE	15 1.40 } 1.55	15.10	1.55	14.50	
10. SPORT	15 2.20 } 2.35	17.45	2.50	17.40	
11. WEATHER	1.15	19.00	1.20	19.00	
12. CLOSING	.35	19.35	.35	19.35	
13. GOODNIGHT/VTR	.50	20.25	.50	20.25	

Example of time chart.

the end. If you are in the situation of having only a few accurate durations when you go on air you cannot therefore work logically from Sequence 1 to Sequence 13 in order to keep a check on the 'on air' running time of the programme. So you work backwards, knowing that you must go into the last fixed items of your programme at specific times.

This tells us that, in order to bring the programme out at 20.25'' we should come to the end of Sequence 7 at 10.50.

Forward timing

If you have a programme which is in two distinct halves, the first being a mixture of live and pre-recorded, the second being a pre-recorded package of fixed duration, you would probably find that you would only forward time to the start of the second half of the programme. Conversely, if the entire first half of your programme is pre-recorded and the second half a mixture, you would probably only backtime, as you would naturally come out of the first half of the programme at the set time.

Buffer

If you have a buffer item of, say, an interview, somewhere in the middle of the programme, you might find it useful to backtime to the end of the interview and forward time to the start of the interview. The difference would give the duration of the interview.

Opting

If your programme has items from other studio centres being transmitted in different parts of the country, then the moment of opting in or out is critical and it is usual to forward time to the opt and back time to the end of the opt.

Commercial break

Again, if you have a commercial break at a fixed point in the programme, you might forward and back time to that point. But generally speaking, the commercial breaks are variable to a certain extent and liaison with presentation would take place about their positioning in the programme.

Critical timings

There will be one or two timings on your time chart which are vitally important. No matter what time you go into a buffer item, you must come out of it at the time specified on your chart in order to bring the programme out to time. If the first buffer item, say an interview, over-runs, you will have to adjust the timing of the next buffer while on air. Circle those critical times in red on your chart so that they stand out.

Writing on script

A time chart is a very good way of seeing the overall shape of the programme and making clear comparisons between durations both before and during the programme.

However, some PAs do not use their time chart when they go into the studio, but transfer their timings on to their script. They do this because it is simpler in the studio to look at just one piece of paper rather than two or three.

Calculating time

Adding and subtracting time rapidly and accurately is difficult. How often PAs have wished that time could be decimalized! Do check and recheck your sums as it is so easy to make mistakes. A stopwatch that also calculates time is invaluable if you can get one.

Don't try to add up whole columns of figures at once. It is far easier to block numbers in groups of three or four. For example:

3.40	1.20	0.30
2.20	3.55	1.25
1.55	2.35	1.25
7.55	7.50	2.10

But, as with everything, you will get faster with practice.

Alternatively, you can round up or down to digits of 5 which are easier to work with. If you have a duration of 11.01″ you could call it 11.00″. A duration of 13.19″ would then become 13.20″ and the second you have lost on the first duration, you would have replaced on the second.

Lack of time!

It is of course very easy to imagine a fictitious programme and work through the different stages. As any PA who has worked on live programmes knows only too well that the luxury of being able to complete time charts and work out, before going on air, the many sums necessary, is minimal, and the likelihood of the running order being adhered to is likewise equally minimal.

12 Marking up scripts

Cameras

Just before the recording you might find it helpful to mark the cameras in large red figures down your camera script. This will be an aid to you when shot calling.

You might also find it helpful to circle the technical instructions in red and note down any standbys or cues you have to give.

News scripts

Some PAs write very little down on their scripts, keeping most of the information in their heads and trusting in their memories and their ability to make instant calculations from a glance at their stopwatch. They work in this way firstly because there might well not be time to do anything else, secondly because once something is written down it tends to become fixed and rigid and difficult to change when changes occur, thirdly because it is time-consuming to keep writing and fourthly because constant alterations will make the script messy and unreadable.

At the other end of the scale is the PA who trusts nothing to her memory, least of all numbers on a stopwatch. Everything will be written down, probably in red pen – all the standbys, all the countdowns, everything.

Most PAs fall somewhere in the middle, probably marking their scripts up as much as possible in the time given, possibly using pencil to start with and only using a red pen in the studio.

'Roll' cues

If the PA is expected to roll the various sources then, whatever else she marks or leaves off her script, she should note the cue to roll.

The operator of the VTR channel will have been alerted by the PA's standby warning and will be waiting, finger poised on the button, for the command to 'Roll!' (or 'Run!' depending upon which term is in common use by the company).

All machines take a certain amount of time to run up to a stable speed, especially if the insert begins with music. Almost all VTR machines are stable after 5″. Some companies run Betacam SP on what they refer to as an 'instant' start, but in effect this is normally a 1″ cue. This is never a good idea if the item starts with music.

Throughout this book I have used a 5″ run-up time for VTR.

Marking the cue

If the presenter's link allows sufficient time, you can mark the script on a word cue. It is standard practice to count three words to a second when working out cues. Therefore when marking up a script for a 5″ roll, you should count back fifteen words in the preceding link.

The speed of delivery might vary from presenter to presenter and you might have to adjust your word cue accordingly, but the difference will be minimal and it is safe to consider three words to a second as the norm.

Ring the precise word, as shown in the example below, and write the cue in large letters – perhaps in pencil to begin with and then confirmed in red pen if you have the chance of a rehearsal and are certain that the cue is correct.

Back to back roll cue

If you are rolling in one source after another in quick succession or have insufficient time in the link to use a word cue, you will need to calculate the time at which you need to roll the second and subsequent sources and mark them on your script.

For example, your first roll into an insert is off a word cue. This item is 30″ long. There is no link into the second insert which lasts 15″ and from this you are straight into the third insert lasting 55″.

You would mark your script first to give a word cue into the first item. Then you would mark down that at 25″ into the first insert you would roll the second, and at 10″ into the second insert you would roll the third.

Channel number

If you know the number of the source channel, assuming that there is more than one, mark that on the script.

Some companies, using a large number of channels, identifies each one by the story title, rather than VTR A, B, C, or 1, 2, 3 etc. Therefore the cue would be 'Roll VTR – Lost Children', rather than 'Roll VTR-12'.

Some or all of the following can be marked on the script, firstly if there is time, secondly if the information is available and thirdly, and most important, if the PA finds it helpful.

Standbys

Standbys are reminders to the PA to give warnings to different people at different times in the course of the programme. Operators of VTR, caption generators, music or tape inserts, presenters, camera operators, anyone in fact directly involved with the programme might need to receive a warning standby from the studio gallery.

SEQUENCE 9 : HIDDEN TREASURE

/S/B VT/

CAM 2 IN VISION	ANDREW Hidden treasure. The very thought of it conjures up visions of gold coins, silver tankards, the rich and precious heritage of days long past. But hidden treasure can be of a very different kind as Mrs Mary Watkins found when she sent off a parcel of old clothes to the Red Cross in response to the latest aid appeal. Chris Johns reports:
Roll VT	
VTR : HIDDEN TREASURE	Dur: 1'35
CAPGEN : Mary Watkins at 12"	Out words : "rummaging in your attics"
CAMS 2 & 3	/Sport link next/
S/B VT	

Marked up script.

Capgens

If name superimpositions are cued in by the PA then they should be marked on the script.

SEQUENCE 1 : TITLES

/S/B VT/

VTR 1 / -5" ROLL VTR-1
TITLES +7" " " 3
+11" " " 1
AT 12" WIPE TO FIRST HEADLINES +14" " " 10
VISUAL +17" " " 6

VTR 3 / Q CHERRY (V/O)
DEVASTATED HOME + FAMILY

A family grieves in the aftermath of today's horrific bomb blast.

VTR 1 /
HOSPITAL

Q ANDREW (V/O)
A new breakthrough in the fight against cancer.

VTR 10 /
RACE COURSE

Q CHERRY (V/O)
The signs are bright for tomorrows big race.

VTR 6 /
ELEPHANT AT ZOO

Q ANDREW (V/O)
And Nelly the elephant has a good night's rest.

CAM 1 /LINK TO BOMB BLAST NEXT/
CHERRY IN VISION

Countdowns into programme.

Countdowns

Some PAs like to write down the countdown timings for each separate item.

In a VTR insert, for example, with an overall duration of 1.25″, you would therefore write the following:

at 25″ : 1.00″ left on VTR
at 55″ : 30″ left on VTR
at 1.10″ : 15″ left on VTR
at 1.15″ : 10″ count out

But there is rarely time on a live news-type programme to work out and write down these figures in advance and there is the danger that, by doing so, the PA will never learn to read her stopwatch quickly and therefore never gain real expertise in live galleries. There is also the tendency, if the numbers are written on the script, of working too closely from the script and not watching the preview monitors.

How to read off your stopwatch without having to write everything down is given in Part Three chapter seven.

But if, by writing it down, it gives the PA the confidence and security to cope with the pressures of the studio at least in the early stages, then the method has its uses.

Counting in and out of programmes

Because of the complexity of many opening title sequences, PAs will usually find that they have to write down the countdowns into the programme involving, as they frequently do, a number of sources to be rolled almost simultaneously (back to back rolls), insert rolls, capgen counts and maybe grams as well.

And when coming off air, all PAs will be meticulous in counting out of the programme and will almost certainly work out and mark up the figures on their scripts, often backtiming from the real, i.e. the 'off air' time.

13 Location work

In the early days of television, programmes were made electronically in the studio, electronically outside the studio, hence the term 'outside broadcast' (OB), or on film. Outside broadcasts initially were always live. With the advent of videotape recording it became possible to relay the pictures back to base and record them for subsequent transmission, possibly after editing. The logical extension of that was to mount a studio VTR in a truck and record the pictures on site. At this point the operation could more accurately be described as an 'outside recording' because the pictures were recorded on location rather than being transmitted back to base. However the term 'outside broadcast' or 'OB' stuck.

No-one would normally call a single electronic camera working into its own VTR, e.g. Beta SP, a single-camera outside broadcast, as it is carrying out the same function as a film camera. However, a similar camera providing a *live* inject into a news programme of, say, a by-election result, *is* referred to as a single-camera OB.

As far as the PA is concerned, location work tends to range from:

- single camera shooting either on film or videotape for something simple, like a documentary, involving herself, a director and a film or video crew,
- single camera shooting on film or videotape for something complex, like a drama, involving costume, make-up, set design, props, artists and an augmented crew,
- multi-camera outside broadcast which can either be recorded on site or transmitted live.

Preparing for work on location follows much the same pattern. Important considerations such as scheduling, travel, food and accommodation are paramount.

Single-camera shooting on videotape

The past few years have seen a revolution in the development and use of lightweight, portable, electronic cameras and this fact, together with the advances made in the field of videotape editing, has meant that more and more programmes which used to be made on film are now being recorded on videotape.

The trend among directors is to utilize the greater freedom offered by these developments by shooting with a film-style technique: using a single camera and recording out of sequence. This method of shooting tends to combine the flexibility of the film-style with the economy and speed of videotape.

Preparations for shooting

The script

If there is a script it will have to be typed, either by yourself or the production secretary. There will be no camera script in single camera shooting but the director might require a shooting script to be typed out and distributed.

Meetings

Depending upon the complexity of the shooting, there may be planning meetings of a technical nature, and meetings with the costume and make-up supervisors. Following these, you will be required to book various facilities, as explained in earlier chapters.

Artists and extras

If your programme requires artists and extras, these have to be booked.

Rehearsals

There might be a period of rehearsal before shooting, in which case you will have to book rehearsal rooms and send letters giving times and places to the artists.

Locations

Locations will be found by the director. On a complex shoot, there may well be a location manager and production manager who will undertake this job. Permission to shoot should be obtained and payments negotiated. If the PA is responsible for this, don't forget that every location requires permission and clearance from *someone*. If shooting is in a public place, you must inform the police. Arrangements should be made for adequate parking close to the location.

Travel

You will have to arrange transport, book hire cars and generally ensure that you know precisely how everyone is getting to the location.

Overseas shooting

If shooting is to take place overseas you should ensure that all travel arrangements are made in good time as air and shipping lines tend to get booked up. Check passports, permits and visas. Contact the press office of the Embassy or High Commission of the country in which you are to work. Arrange the necessary travellers cheques and local currency.

You will need to ensure that correct vaccinations or inoculations are obtained for all the crew if these are needed and take out health and personal effects insurance for the production crew.

If a British crew is booked, an important part of the PA's job will be in the preparation of the 'carnet'. This is a detailed list for H.M. Customs of all stock and equipment being taken abroad. A number of copies are required and pro forma invoices for consumables, i.e. tapes and batteries, are also necessary.

When booking flights, bear in mind that you will probably be carrying a good deal of excess baggage. Sometimes deals can be arranged with airlines. It might be necessary to hire a shipping agent to meet you and help clearance through Customs.

'Would you be prepared to give them yours?'

Accommodation

You will need to book suitable accommodation for the unit and that can be a difficult business as everyone will have different expectations and requirements. Do not forget that the more remote locations will probably have a scarcity of decent hotels and you should book as early as possible. Do not wait until you have the names of all the artists – block book rooms and send a detailed list later. Remember that it is better to overbook than be faced with irate artists or members of the production team who have no bed for the night. Would you be prepared to give them yours? And remember that you will never, ever satisfy every single person on the unit, no matter how hard you try.

Food

Neither will you satisfy everyone when it comes to making arrangements for eating. Faced with a gourmet's meal of smoked salmon, caviar and tempting home-made dishes there will be members of the unit who will slink off to the nearest cafe for fish and chips.

If the unit is a large one, it would be best for you to engage professional location caterers rather than rely on local restaurants. If the unit is small it might become your responsibility to find suitable eating places at each location, and don't forget the growing number of vegetarians.

Petty cash

On a small production you might be required to hold the petty cash, to pay for meals and accommodation and any incidental expenses incurred. This could lead to you having to carry vast sums of money around which would be a constant worry. Wherever possible you should arrange banking facilities in this country or overseas.

Schedule of shooting

Sometimes the schedule of shooting is worked out by the production manager, if there is one, and sometimes by the PA in conjunction with the director. The schedule is organized with regard to the different locations, the amount of shooting at each location, the availability of artists and so on. Time should be allowed for travel from one location to the next and the regulations governing hours of work must be adhered to. When the schedule has been finalized it should be typed out and distributed. As well as the details of the scenes to be recorded each day it should contain:

1. A cast list, together with agents' telephone numbers (artists' home numbers should be restricted to the immediate production team).
2. Detailed information on travel arrangements with maps to assist those driving to the locations.

3. A list of accommodation booked, the address and telephone number of the hotel, the dates and names of those booked in.
4. Details of the locations with addresses and telephone numbers of contacts.
5. Any other useful addresses or telephone numbers, i.e. local taxi service, doctor, police and so on.
6. Details of the technical equipment required, special props etc.

Outside broadcasts

When television was still in its infancy, the outside broadcast represented one of the most exciting areas of the whole exciting new medium. The fact that the viewer could see the Cup Final of the football match, the grand State occasion, the gala variety show, as it was actually happening and usually from a better vantage point than the spectators at the event was, and still is, what television was all about for many people.

With the development of cable and satellite, the audience for these live events became world-wide. On July 29, 1981 when His Royal Highness Prince Charles married Lady Diana Spencer in St. Paul's Cathedral, London, live pictures were relayed to more than seventy countries, enabling over one eighth of the world's population to share in the occasion as it happened. On Saturday July 13, 1985 the Live Aid concerts at Wembley and Philadelphia in aid of famine relief involved eight satellites and reached an estimated audience of 1.5 billion – watching eighty-five percent of the world's television sets. For the Production Assistants sitting in the mobile control rooms at the heart of these massive spiders' webs of communications, these events must have posed awesome responsibilities.

Involvement in programme content

Unlike the live news or magazine format programme where the PA's role is often restricted to one of time-keeping, on a live OB the PA is often closely involved not only in the research but also in the creative content of the programme.

Frequently there is only a producer (or director) and PA working on the programme and the PA is thus involved from the early setting up stages.

Preparing for the OB

The preparatory stage of a live OB involves the PA in a range of duties similar to the setting up of documentary filming. Dealing with people, making contacts, booking facilities, arranging transport and accommodation, liaison with the police, local authorities and countless other official bodies, getting archive material, attending planning meetings, compiling and typing out

'The PA should familiarize herself as much as possible with what is to happen during the OB.'

planning sheets, schedules, technical requirement details, the script and/or running orders are all standard tasks, many of which have already been gone into in earlier sections of this book.

What is of paramount importance however is that the PA should familiarize herself as much as possible with what is to happen during the OB. This is because on many OBs there is one job of overriding importance for the PA – a job known as 'living in the future'.

Living in the future

On a live outside broadcast which has no script and is largely dictated by the event itself, the PA's prime contribution is to provide a running commentary of the event, not by describing what is happening at that moment but by talking through what is to come.

The PA is able to provide this commentary by means of her detailed and intimate knowledge of the event. It means that she needs to have, not just a vague idea of the programme content but extremely well researched and memorized information.

For example, in an OB containing a ceremonial procession, the PA should know the route of the march, the order of the procession, the dignitaries involved. She should be able to put names to faces and recognize rank and title. She needs to be able to inform the director that 'X's car has just left the Town Hall' or that 'Y is just coming into shot on camera 5' with a certainty that it *is* X and Y and not A and B!

She needs, in fact, to learn the story of the event: the rules and regulations of each sport, the routine of the air display, the seating arrangements at St. Paul's Cathedral etc. She needs to have attended rehearsals, on-site recces and planning meetings. The more information she has absorbed beforehand, the more valuable her contribution will be during transmission.

Part Three: Production

1 The studio recording

What to take to the studio

It is generally the PA's lot to go from place to place burdened like a packhorse and for the studio recording it is necessary to virtually transport your entire production office and set up home in the confined space of the studio gallery.

You should take with you:

Scripts

If there has not been time to distribute these in advance, then you must do so as first priority on the morning of the studio.

You should give copies to those in the gallery – the vision mixer, the studio supervisor, the capgen operator (if you have one). Copies should be left in the sound and lighting control areas.

You should leave copies at strategic points on the studio floor, give a copy to VTR where the programme is to be recorded. Leave copies in the Operations Rooms. Any remaining copies should be kept with you in case of need.

Recording orders/running orders

Spare recording orders/running orders should be left in the studio and in the gallery.

Studio call sheets

Should be typed up with the artists' names then given to the stage manager who will fill in the call times.

Rehearsal scripts

You might like to take a few copies of the rehearsal script to the studio. This is not applicable to the live or as-live programme.

Camera cards

These are to be distributed to the camera operators first thing. You might not have camera cards for a live or as-live programme.

Programme file

Despite its weight you might find it useful to take with you to the studio. If you leave it in the office you will undoubtedly need it.

Address and telephone list

Anyone in any way connected with the production should be included in your address and telephone list and you should take it to the studio.

Key personnel in the building

The telephone numbers of the key people in the building, i.e. reception, canteen, security etc. should be to hand.

Time chart

You should either take your own rehearsal script with the written-in timings taken from the final rehearsal, or have a time chart made out with these timings for each individual scene, a column for cumulative timings and several blank columns.

For a live or as-live programme, take the time chart you have been working on during the run-up to the studio. It will contain the estimated timings, the real timings, where known, the cumulative timings and perhaps the start of backtiming.

Records/tapes, CDs

You should ensure that any music needed for the studio is taken to the sound control areas. Likewise any sound effects on disc or tape, although these will probably have been acquired by the sound supervisor.

Stopwatch(es)

Make sure you have more than one stopwatch with you and ensure that they are wound up and in good working order.

Pencils/coloured pens/erasers/ruler

All necessary office equipment should be taken to the gallery, including pencil sharpener, hole puncher, stapler, paperclips and scissors.

Blank and ruled paper

Take sufficient with you for taking down notes from the director and notes for editing.

Throat pastilles

You can get quite hoarse after a twelve hour studio so take some soothing throat sweets.

And finally most PAs carry a selection of the following at the bottom of their seemingly bottomless bags – aspirins, indigestion tablets, tissues, safety pins . . . the list can be endless.

The studio floor

You arrive in the gallery at an early hour, deposit your travelling office on the desk and settle down to . . .

Not quite yet. Before making yourself comfortable in the gallery you must go round distributing camera cards, running orders, scripts and so on, and in so doing you should get to know the people involved in your production. For the rest of the day – or days – you and they will be simply disembodied voices. It is far better for you to meet face to face. It will make for a better working atmosphere and you will not have time later on.

Some camera operators have commented that many PAs seem scared of going down onto the studio floor. They act as if they are trespassing and approach people in a furtive manner, heads down, silently handing over the camera cards before scuttling back to the relative security of the gallery. Whether this is due to an inbuilt and unfounded sense of superiority or, more likely, a strong sense of inferiority, of feeling that one knows one's place as PA and that place is *not* on the floor of the studio is unclear, but camera operators are known to be human and, like everyone else, they want a PA to whom they can relate, a real person as opposed to a voice.

So do go down to the studio, walk round the sets – taking care of course not to get in the way – find out the names of the camera operators and boom swingers and become aware that you belong here as well as in the gallery.

To help you do this, let us wander round the floor of a fictitious drama.

The sets

These will have been constructed overnight in all probability. Their design will not be entirely unfamiliar to you as you will have seen the scale model and floor plan and watched the final rehearsals. However it will look different from your own conception if only because of the addition of props and furnishings.

The designer, the stage manager and a host of stage hands will be busy on the sets, adding the final touches and sorting out any last minute problems.

Lighting

The lights will be high above you, bank upon bank of them. Electricians will be positioning them and they will be controlled by the lighting control area adjacent to the main gallery.

Cameras

Studio cameras give high quality pictures. They are fitted onto different sorts of mountings according to the needs of the production. Full communication is standard enabling talkback to and from the director and PA.

In addition, the sort of lightweight camera used on single camera location shooting is increasingly being used in studio productions. These will have full talkback facilities and may be fitted with a larger viewfinder than when used on location. Lightweight cameras might well be used hand held but they require two people to each camera – an operator and someone to guide the cable.

Mini-cams which are just two inches long might be used to give another output. These can be clipped-on to any object, such as the end of a guitar, or over a frying pan, to give unusual static shots. They do not require an operator, but a cameraman will be needed to rig the mini-cam.

Camera mountings

Pedestals (peds) are the bread-and-butter of most studio productions. Cameras are mounted on the pedestals which give them a great amount of stability. The pedestals are on wheels so they can be used for crabbing and tracking shots. They are extremely versatile since they can also vary the height of the camera.

Cranes are mobile camera mountings which can be raised and lowered to a greater degree than the pedestal. They can be used very effectively in a production but require three people to operate them.

A standard studio might come equipped with four peds and two hand held cameras.

Sound

Microphones would be placed at the positions laid down in the studio floor plan. They are held in place over the sets by means of booms, which are telescopic arms. Each boom has its individual operator.

'So go down to the floor of the studio, walk around the sets – taking care not to get in the way . . .'

The gallery

If you feel that you are only allowed in the studio on sufferance then your return to the gallery will be like a home-coming. But this security will only be truly felt after your belongings have been spread out to your satisfaction and your base established.

Every PA will have her own precise, rigid way of setting out her script, blank paper, ruler, pens, pencils, stopwatches and so on. She will arrange them in a pattern most pleasing to herself and for her greatest comfort and accessibility. She needs to be reassured in advance that if she stretches out her hand for a stopwatch, a pencil, a ruler, that object will be waiting at the precise spot without her having to hunt for it on the desk.

The control desk

People

Your seat is at the side of the director. On your other side will, in all probability, sit the producer. Also at the desk will be the vision mixer – unless the director performs that operation for himself – possibly an effects operator, a capgen operator and the studio supervisor.

Equipment

The desk will contain a number of keys and buttons, including the key to activate the red warning transmission light, telephones, a microphone on a stalk, possibly an insert stopwatch or the real time displayed digitally and maybe a digital countdown clock.

Directly in front of the desk is a row of monitors of different sizes, the smaller ones showing the output of each individual source, i.e. the cameras, VTR, capgens; the larger ones being the preview and transmission monitors.

Timecode will be displayed on the recording VTR's monitor and there will be the studio clock, hopefully in a clear position.

Communication

Communication between the gallery and the studio floor is by means of a talkback system which is not picked up by the studio microphones.

Open talkback

Open talkback means that everything spoken in the gallery is heard by the presenter, floor manager, camera operators and anyone else in the studio who is using either an ear-piece or headset (known as cans). It is also heard by all incoming sources such as the videotape operators.

Switched talkback

Switched talkback means that the presenter will only hear the gallery when a switch is held down by the PA or director. This is less distracting for them than hearing everything that is said in the gallery. However many presenters choose to stay on open talkback.

2 Gallery duties

Depending upon the content and type of programme that is to be recorded or transmitted, the PA will have different specific duties. But some jobs are general to all gallery work:

Central point of communication

The PA's overriding task in a studio gallery is to act as the focus of communication by keeping everyone informed verbally of what is happening at the moment and what is to come next. It is an important task because most of the technical people on the production will have neither the time nor the facility for keeping their eyes on the script or running order. If there is no script, the importance of this job is self-evident.

On a live, news-type programme, the PA will perform this function partly by talking the studio through the running order, partly by counting through the items and partly by keeping the studio constantly informed as to how much time is left.

On an unscripted outside broadcast which is dictated by the events that are being transmitted, the PA provides an essential commentary on what is happening now and what is about to happen by means of 'living in the future'.

On a scripted programme, whether live or recorded, the PA performs the same task in a very precise and exact fashion by means of shot calling.

Answering queries

There will be a lot of queries arising from all areas concerning, for example, equipment that has not arrived, organizational matters that need attending to, anything and everything. Deal with the matters you know about yourself and pass others on to the right person. Part of your job is to act as a sorting house, routing queries in the right direction.

Answering the telephone

If the telephone needs answering it will not ring but a white light will flash. You should answer it during rehearsals, unless you are otherwise engaged when the studio supervisor will answer it. You will know whether or not the call is urgent.

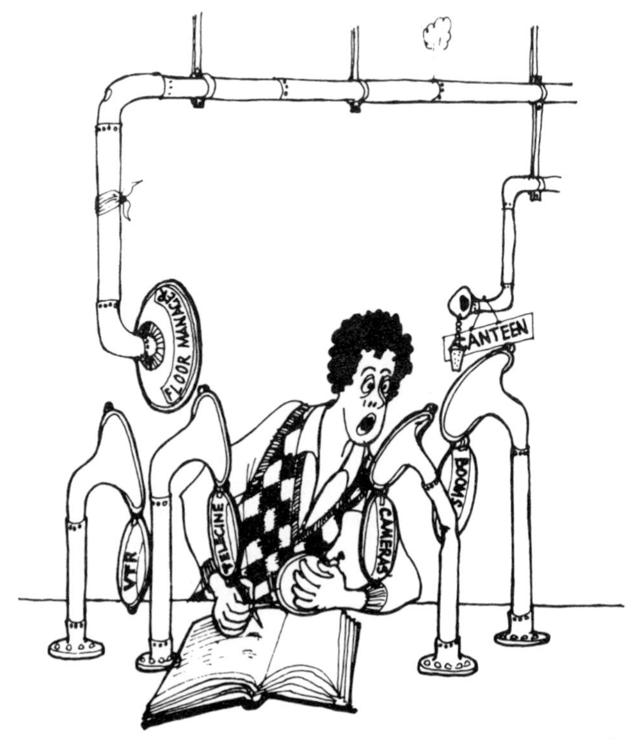

'One of the main duties of the PA when working in the control gallery of a studio is to keep everyone informed verbally of what is happening . . .'

Don't, for example, interrupt the rehearsal to tell the actor Fred Bloggs that his cleaning lady forgot to collect the vacuum cleaner from the repairers and therefore his house will remain dirty. You can tell him later. If agents ring, generally you can pass the message on at a convenient break in rehearsal. Just use your intelligence.

Camera rehearsal

If a programme is to be recorded, the first thing that happens is a rehearsal with cameras, otherwise known as the 'stagger through'. The director will rehearse one scene, or a section of that scene or a whole piece of the script with the actors and will talk the camera operators through the shots he wants. This rehearsal might go on for a morning, a whole day or, if the production is to be rehearsed then recorded in chunks, there might only be a short rehearsal before recording.

Some directors conduct the rehearsals totally from the gallery, some conduct them from the floor of the studio. If the director takes the rehearsals from the floor you should remain in the gallery to act as general liaison in addition to calling the shots.

As directed

The director will talk through any 'as directed' sections of the script, explaining to the camera operators the framing of the shots required.

News rehearsal

On a live news-type programme, there might not be time to conduct a full rehearsal. In that case there will be a script check where the director will run rapidly through the sequences and iron out any technical problems.

Note taking

The director might expect you to take notes throughout the rehearsal. Most notes will be given as muttered asides and sometimes you will not be sure whether you are meant to write them down. Do note down everything.

Group the notes under headings, i.e. notes for artists, notes for sound and so on as this will save time later when the director gives them out.

If the director has been rehearsing from the gallery he will usually go down to the studio floor once the rehearsal is over. You will accompany him armed with the notes. Either you will be asked to give them directly to people or, more likely, you will be asked to remind him of the points he wishes to raise.

If you have been using shorthand to take down the notes do make sure that you can read them back quickly and accurately.

Script changes

You should mark any script changes that occur as rehearsals progress whether these changes are in the dialogue or concerning the cameras. As the director sees the shots he will undoubtedly wish to make revisions to the script, to add shots, develop shots, change camera positions and so on. You must ensure that everyone is aware of these changes, if necessary by talking them through the script at the end of rehearsal and prior to recording.

Throughout the studio your script will probably get messy as things change for a second and even third time. You might well be tempted to mark up a fresh script just before the recording. If you do you will almost certainly leave out some vital alteration or instruction. So don't do it! It is far better for you to take plenty of erasers and rub out unwanted instructions as you work through the rehearsals.

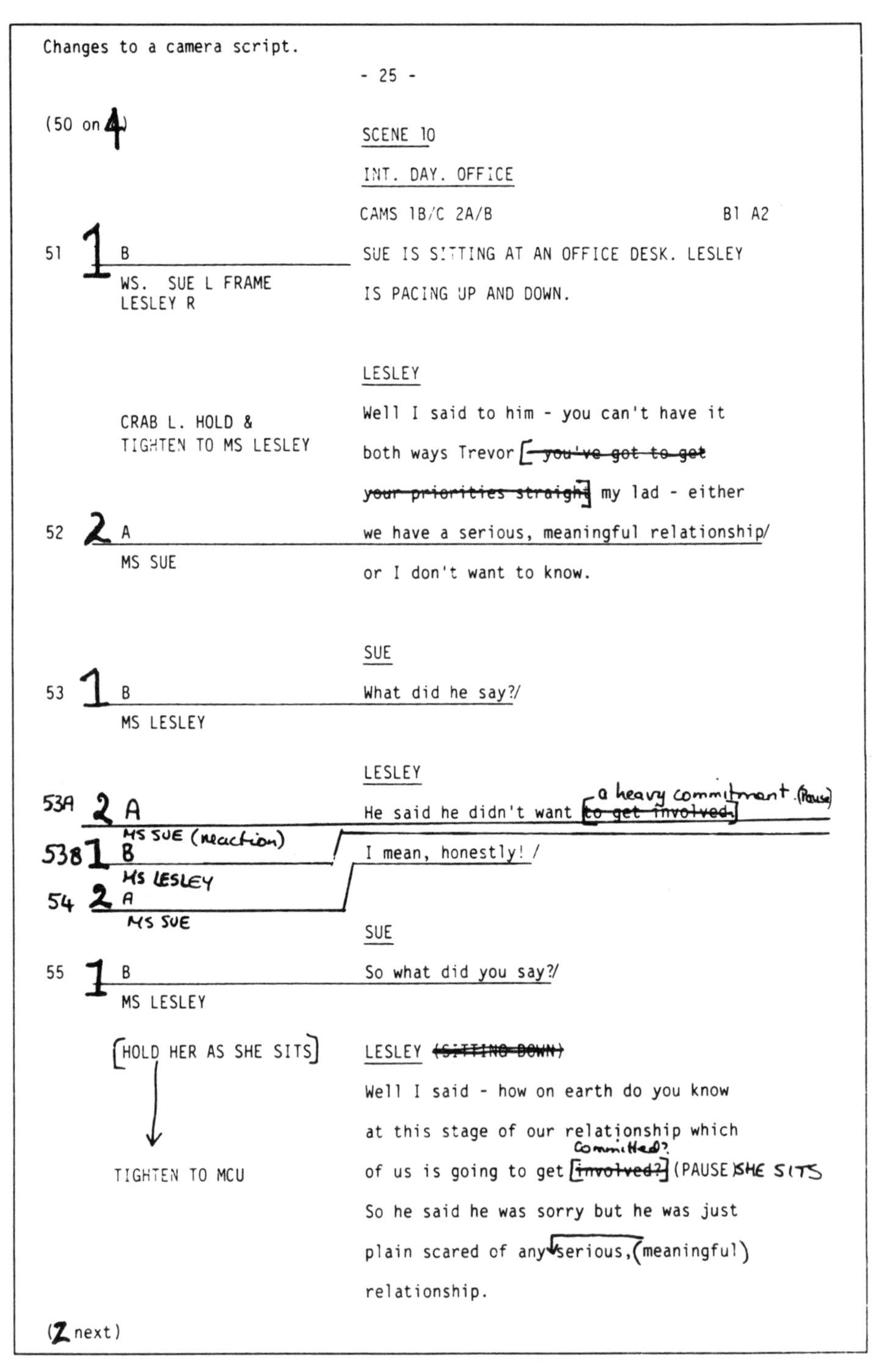
Changes to a camera script.

- 25 -

(50 on 4)

SCENE 10

INT. DAY. OFFICE

CAMS 1B/C 2A/B B1 A2

51 1 B
WS. SUE L FRAME
LESLEY R

SUE IS SITTING AT AN OFFICE DESK. LESLEY IS PACING UP AND DOWN.

LESLEY

CRAB L. HOLD &
TIGHTEN TO MS LESLEY

Well I said to him - you can't have it both ways Trevor [~~you've got to get your priorities straight~~] my lad - either

52 2 A
MS SUE

we have a serious, meaningful relationship/ or I don't want to know.

SUE

53 1 B
MS LESLEY

What did he say?/

LESLEY

53A 2 A
MS SUE (reaction)

He said he didn't want [~~to get involved~~] a heavy commitment (Pause)

53B 1 B
MS LESLEY

I mean, honestly! /

54 2 A
MS SUE

SUE

55 1 B
MS LESLEY

So what did you say?/

[HOLD HER AS SHE SITS]

TIGHTEN TO MCU

LESLEY ~~(SITTING DOWN)~~

Well I said - how on earth do you know at this stage of our relationship which of us is going to get [~~involved?~~] committed? (PAUSE) SHE SITS So he said he was sorry but he was just plain scared of any serious, (meaningful) relationship.

(2 next)

Example of changes to a camera script.

Speaking out loud

The first time you have to speak out loud in the gallery can be quite unnerving. There you are, speaking into a microphone at no great distance from your nose, and the tendency will be to mumble in embarrassment. Below are a few dos and don'ts which may be of help.

Dos and don'ts in the gallery

Whether the atmosphere in the studio is relaxed and good-humoured or nail-bitingly tense, there are certain ways of speaking and acting as a PA which should be adhered to. These are disciplines of speech and manner which should be learnt as they make for a professional and smooth running studio.

1. Do not speak unless you have something to say which is relevant to the programme. Some directors insist on a quiet and concentrated atmosphere while others favour a more casual approach. Whatever the individual director's style, you should remember that it is very confusing for everyone listening in to hear a lot of extraneous talk. In addition, the more your voice is heard talking about unimportant, unrelated matters, the less likely it is that people will take notice of you when you do say something pertinent and important.
2. When you need to speak, speak clearly and confidently. If you are too quiet you will not be heard and if you are too hesitant your instructions will lack conviction.
3. Always preface your remarks to the person intended to hear them. 'Fred on grams – would you pre-face the music 5″ earlier', alerts Fred that you have something to say to him and he is then ready to take in your instruction. It also tells everyone else that the message is not for them. If you say, 'Pre-fade the music 5″ earlier would you Fred', Fred might not have realized that the message was intended for him until too late and everyone else would have had to have listened to the whole message before realizing that it was not meant for them. After a bit they will cease to listen at all.
4. When giving a general warning you should repeat the vital part if there is time. i.e. 'One minute to transmission, one minute'. It gives people who are busy with other things a second chance to hear your warning and fixes it firmly in their heads.
5. It is courteous to get to know the names of the camera operators and, indeed, everyone associated with the programme.
6. Try not to talk over the director. It can be difficult if a director is giving a long, complex note to, say, the floor manager, and you have an urgent message, but it is rude to butt in; it will confuse those listening, especially the floor manager who will probably forget what the director has told him; it will infuriate the director and make for a very uncomfortable

relationship. At least wait for the director to take a breath before you jump in.

7. You must always keep the studio informed about what is to happen next. If it is to be a recording break, a VTR insert, a tape run-on while a camera is re-positioned, let everyone know. Each individual shot or sequence leads on to something. At the end it leads to the end of the programme or coming off air and you should say so.
8. During a long shot on one camera, do remind everyone what is happening, i.e. 'still on shot 51 on 5' and issue a further warning to the next camera operator near the end of the shot, i.e. 'coming to 2'.
9 The director and vision mixer will normally remind camera operators of moves, but if they do not then you should. For example, if, during a sequence intercutting between cameras 1 and 2 (shots 31–40), camera 3 is required to move to position B to be ready for shot 41, just ensure that camera 3 *has* moved and *is* in position. It is a bit late to warn him if you have already reached shot 40 without him moving.

'If you wish to warn the camera operator about a move or a change or communicate in any way, do be polite about it and don't shout instructions.'

10. At all times, if you wish to warn the camera operator about amove or a change or communicate with any of the studio staff in any way, do be polite about it and don't shout instructions. If the camera operator wishes to communicate with the gallery he might well move his camera vigorously from side to side or up and down. If you see this movement, tell the director.
11. Whatever else you are engaged on, always keep an ear open for what the director is saying and for what is happening around you. Don't get so involved in your calculations that you are oblivious to everything else.
12. The situation can get pretty tense in the gallery and people can say things in the heat of the moment they would afterwards regret. Be sure you never do so. Never, never swear. It can be understandable in certain conditions but it is not clever, it will not gain you the respect of the studio and my advice is to leave it to others. Yours is the still, cool, calm voice in the midst of whatever furore is happening around you.
13. Try to keep a sense of proportion about the job. There *is* life outside the studio and if you can retain your sense of proportion and your sense of humour you will find it easier to cope. Don't forget that if you are finding yourself under a lot of pressure, so is the director.
14. Finally, if you are in the wrong, don't try to cover up or shelve responsibility on others. Too many people try to do that. Always admit your mistake and apologise.

'Always admit your mistakes . . .'

Keep your head up! The importance of previewing

Having started my broadcasting career in radio, not television, I was used, when recording a radio show, to keep my eyes firmly fixed on my stopwatch and script on the desk in front of me. There was no point or necessity for me to watch the performers in the studio grouped, as they were, around a microphone.

When I began work in television I found it extremely difficult to tear my eyes away from the script and stopwatch in order to look up at the monitors. There was, I think, an innate fear that by raising my head I would lose my place in my script, be unable to read my stopwatch and lose my grip on the programme.

I know that I am not alone in that fear even among PAs who did not start, as I did, in radio. There are many PAs who find it very hard to look up from their watches and scripts. It is, in one sense, a striving after security. The script is simple enough to relate to and understand; the stopwatch likewise. But the moment you start watching the many different monitors there is the danger of becoming confused.

On any television programme however it is essential to watch the preview monitor for what is coming next.

Importance when shot calling

On a programme where your main task is to shot call, it seems very easy just to look at your script, listen to the dialogue and call the shots according to the scripted cuts. If you do that, however, you will probably go wrong and once having gone wrong you will continue to call the wrong shots throughout the sequence unless you check the monitor.

By being inaccurate in shot calling you could easily confuse the camera operators as well as everyone else listening to you. However, everyone would soon realize that you were unreliable and would cease to listen. You *must* be accurate or not speak at all. And you can *only* ensure continuing accuracy if you keep your head up and watch the transmission monitor. It is only after the shot has been punched up on that monitor that you should speak.

You should also check the monitor of the camera that is to come next. If the camera operator is not ready with the required shot you might need to give a further reminder, i.e. 'Coming to *you* next, 2'.

You might find it helpful to hold your script on a level with the monitors so that your eyes can flick comfortably from one to the other rather than having to raise and lower your head each time you call a shot.

Importance during live and as-live programmes

On a live or as-live programme it is vital for you to watch the monitors for the following reasons:

Watching the ident

When giving a standby warning for an insert item, you should always check that the correct insert is ready on the correct channel. There will be a visual ident on an item on VTR in the form of a clock and the title of the piece.

It is not enough to assume that the right stories will automatically be ready on the right machines. In nine times out of ten they will be, but the make-up, as it is called, of which items go to which channels can sometimes be very

'And it has happened!'

complicated and mistakes can happen especially if one or more of the stories arrive late or the running order changes at the last minute. You are the last check before transmission that the correct item will follow the presenter's link.

Counting in from the ident

You should always count into the insert from the ident clock, while watching the monitor, and not from your watch. You should start your insert watch from the moment zero is reached. This will give you an accurate timing for the item.

Counting out of insert

You will look a complete fool and be responsible for confusing the entire studio if you count out of an item without watching the transmission monitor. You might have been given a wrong duration or the director might decide to come out of the insert item early and there you would be, your eyes glued to your stopwatch, counting out of an item that possibly finished five seconds ago, or that continues long after you have supposedly counted it out. In addition, if you are counting out of an item that has already finished, your overall timing will be inaccurate.

Do beware of the pre-recorded item that contains shots of the presenter in the studio. It is very easy to panic and think that you are out of the recording and live in the studio.

The ultimate in counting out would be to count out of the entire programme oblivious of the fact that it came off the air ten seconds ago. And it has happened!

Cueing in capgens

On your source sheet you have made a note to cue in the name super of Mrs Joan Bloggs, Chairman of the Blogsworth District Council at 1.25″ into the item. At 1.20″ into the item therefore you count in to the super, the vision mixer cuts it up or fades it in, and it appears superimposed not over a shot of the honourable lady but over a shot of a large pig! Neither you nor the vision mixer were watching the monitor.

Previewing is very hard for a number of PAs, especially those involved with live programmes where the stopwatch can become an obsession. But it is essential. And for those people who say that it is just not possible to keep an intelligent eye on the monitors while working from the script and stopwatches, the answer must be to practise and work out for yourselves how best you can accommodate these things.

3 Shot calling

I have already stated that the PA is the central point of communications, both in the setting up and during the recording or transmission of a programme.

On a closely scripted programme the PA performs this task in a very precise and exact fashion by means of shot calling.

It is precise because each different shot has been given a consecutive number either by the director when drafting the script or by the PA when typing it in its finished form. By calling out the number of the shot she keeps the entire studio informed about the place which has been reached in the script.

Warning for camera operators

In addition to calling out the shot, the PA should also give the camera operators a warning of which camera is to be used next. It must be remembered that the camera operators work from the list of shots relating to their own camera as written on their camera cards. They do not usually work from the full camera script therefore they rely very much on the PA giving them warning standbys.

What to say

As in almost everything in television, there is no standardized method of shot calling. At least two systems are currently in use in the UK:

1. The first system of shot calling is to call the number of the shot that is currently being shown on the transmission monitor and the number of the camera whom the vision mixer will cut to next. For example: 'Shot 40, 2 next' – meaning that we are currently recording or transmitting shot number 40 and the next shot will be on camera 2. '41, 1 next, 42, 2 next, 43, 4 next' and so on.
2. In the second system of shot calling the PA does not call out the number of the shot that is currently being transmitted or recorded but calls out both the shot number and camera that will come next. For example: 'Shot 41 on 2 next' – meaning that the next shot will be number 41 on camera 2, (the understanding being that we are currently on shot 40). '42 on 1 next, 43 on 2 next, 44 on 4 next' and so on.

I cannot say that both systems have their advantages. My feeling is that the first system is far better both for the PA who is calling the shots and for those at the other end of the communications system who are listening.

My reasons are:

- The second method does not fulfil the requirement of telling everyone what is happening *at that moment* as well as what is to happen next, and it must be remembered that camera operators are not the only people who are listening to the PA. Some of the recipients are more interested in the current shot and are not necessarily interested in the one that is to come next.
- By using the second method the PA is working one step ahead of everyone else and it would be relatively easy for her to get confused and start calling the wrong shots. That, for everyone concerned, is far worse than not calling any shots at all.
- On a fast shot calling sequence, the PA using the first method drops the advance warning to camera operators and concentrates on the actual shot that has been punched up on the transmission monitor, i.e. '41, 42, 43, 44' and so on. But if the second method were used the PA would have to add the word 'next' to every single shot she called, i.e. '41 next, 42 next' and so on. While this may seem trivial to anyone who has not had to conduct a fast shot calling sequence, believe me that 'next' would matter and could make it difficult for the PA to keep up with the shot changes.

Having said all that, however, the individual PA will have to use whatever method is employed in the company for which she works.

4 Timing and logging

Timing

Certain aspects of the PA's job have priority over others depending upon the type of programme on which she is working. On a programme which is to be recorded and which has ample editing before transmission, timing is not a major consideration.

On a programme which is recorded with only a minimum of editing, timing becomes more important.

Read-through and final run timings

At the initial read-through and again during the final rehearsal you will have taken timings. Most probably you will have written them on the pages of your rehearsal script, ideally at one minute intervals or at least once on every page of script. The timings will be taken from your stopwatch and you will have added together the durations of each scene until you have an overall timing.

The timings you take at the final rehearsal will be of most use to the director and producer because, as a result of your calculations scenes might have to be dropped or extensive re-writes done with the aim of cutting down or building up the script before going into the studio.

So, hopefully, you will go into the studio with roughly the right length script for the time slot allocated, or at any rate with a good idea of by how much it is likely to over- or under-run.

Don't forget, however, to allow more time if you have a number of action props that have only been mimed in rehearsal. Drawing a cork out of a wine bottle for example might only take seconds when mimed in rehearsal, but takes very much longer with the real thing.

Another consideration when calculating overall times should be that of audience reactions during light entertainment programmes.

Timings during the studio

During the camera rehearsals you should get into the habit of taking timings and noting down the durations of scenes or sections. During the recordings you should switch to red or another colour pen and continue to take down the timings.

	SCENE/SET	REH. TIMES	CUM.	STUDIO TIMES	CUM.
1	DAY. EXT. STREET (FILM)	2.10		(F) 2.10	
8	DAY. EXT. CAR PARK (FILM)	3.40	5.50	(F) 3.40	5.50
14	INT. ENTRANCE	.20	6.10	.20	6.10
15	CORRIDOR	1.00	7.10	1.10	7.20
2	KITCHEN	5.10	12.20	5.05	12.25
4	"	2.15	14.35	2.20	14.45
12	"	4.05	18.40		
18	"	1.00	19.40		
13	DINING ROOM	3.05	22.45		
16	"	6.25	29.10		
20	"	6.00	35.10		
3	LOUNGE	2.00	37.10		
5	"	.35	37.45		
7	"	2.10	39.55		
9	"	1.05	41.00		
17	"	4.15	45.15		
19	"	.15	45.30		
6	BEDROOM	3.10	48.40		
10	"	2.50	51.30		
11	HALLWAY	.10	51.40		

Time chart for a 50-minute studio drama. The scenes are listed in recording order as otherwise it would be impossible to add up a cumulative running time during the recording.

Some PAs note the timing at the foot of each page of script, while others write it at the top of each page or half-way down. If you note the timing half-way down each page you will be less likely to lose your place in the script as you turn over the page. Whatever you do, however, be consistent.

During the recording you should compare the real times with the rehearsal times in order to see whether the programme is over- or under-running.

Remember that you should only time from the moment the director says 'Cue' or 'Action'.

Time code logging

Few programmes nowadays are recorded in their final story order. There are many reasons for this, the most important being the complexity of programmes today allied to the vast technological improvements in videotape editing.

As a result we have scenes or sequences recorded in sections, recorded in an order which is determined by the number and type of sets, the facility of camera movements, the availability of artists and so on. On a multi-episodic series, scenes from different episodes will be recorded on the same studio day, i.e. all the scenes set in the kitchen would be shot together.

Film or video inserts are only very rarely played directly into a studio recording but are usually edited in later.

In addition, directors used to the flexibility of the single camera 'film' style technique of shooting, often adapt it as best they can to the multi-camera situation, re-recording scenes shot multi-camera in order to obtain further shots from different angles.

Then again, each section that is recorded might have a number of takes for all sorts of reasons ranging from inadequate performances by the artists to errors of a technical nature, culminating in the director's 'gut' feeling that the scene could be improved by re-taking it.

For some complex programmes the outputs of some or all of the individual cameras might, in addition, be routed to individual VTRs, thus enabling the programme to be substantially re-edited afterwards. These are known as 'isolated feeds' or 'isos'. For example, a game show might be recorded on two VTRs, a main, which is the composite programme, and an iso of the camera which is framed on the presenter. These two recordings should normally have matching timecodes.

So we end up after the studio with a mass of unrelated material which must now be reassembled, sorted and edited into correct story order.

It is therefore of prime importance that the PA keeps a note of what is recorded and the order in which it has been shot. She needs to note the number of takes, their duration and the reason why they were not considered usable. To do this she needs some means of identifying the position of each recorded section on the rolls of videotape.

In the past she would have done this by using a stopwatch as a means of reference, adding up a cumulative 'tape running' time and noting down her specific times from that. This was not that accurate and could become very complicated. The advent of time code revolutionized the PA's work in this respect by providing an identification which is recorded on every frame of videotape.

What is time code?

Time code was developed in 1967. It is a method whereby each video frame is identified by means of numbers broken down into hours, minutes, seconds and frames. Each frame always keeps its original identity, known as the time code address. This makes logging and editing more precise and efficient.

Time code is normally recorded on an audio track which is then played through a time code reader which produces the relevant numbers. These numbers are often inset in a box, white letters on black, and are displayed on a monitor in the gallery in hours, minutes and seconds (the frame digits are often not displayed in the gallery for simplicity).

Some studio galleries have a time code reader on the desk which the PA can freeze in order to write down the time. When she cancels the button the time code will revert to the present.

The time code recorded on to the tape can either show the tape running time or the actual time of day. Some companies use one system, some another and some a combination of the two depending upon the programme.

Tape running time

This has the advantage of letting the PA know the position reached on the tape and how much blank tape is left. There is also no possibility of duplication of numbers and tape running time code tends to be preferred by videotape editors as time of day time code can cause problems in editing.

Time of day time code

This system records the actual time of day on the tape. It is most useful on location when the PA is trying to compile a shot list without having access to a monitor with inset time code. She can then take the time from a digital watch.

However when the same tape is used two days running there is the possibility of a duplication of time code numbers on the same roll. Either a fresh roll should be started each day or the PA should note down the date as well as the time code.

Recording log

The PA should compile an accurate log of the recording by noting down the 'in' points of time code at the start of each section of recording and noting the information given below.

It is not necessary for her to note down the 'out' points of time code at the moment the recording is stopped unless she is not using a stopwatch and needs to work out the running time of each recorded section solely from the time code readouts.

This log is built up during the recording and written out during the studio. It should contain the following:

Time code logging.

Videotape roll number

This is most important if a number of rolls are used during the course of the studio.

Time code 'in' point

This should be from the moment the director cues the action in the studio.

Shot numbers

Taken from the camera script of the recorded section.

Episode and scene numbers

Should be marked on the log, especially if the recording is multi-episodic, as it is vital to preface every scene with its correct episode.

Take number

This should correspond with the verbal ident given by the floor manager.

Notes

Anything that was wrong with the recorded section should be noted down and any additional notes or reminders for the editor should be written in this column.

VTR ROLL 9 — RECORDING LOG — "THE LONG SEARCH"
REC: 25 September 1985

"In" t/c	Shot nos.	Ep/Scene	Take	Notes	R/T
01.15.26	69-74	3.24	1	Boom in shot	2.00
01.17.30	"	"	2	OK	2.00
01.19.40	75-85	3.37	1	OK	3.20
01.23.10	86	2.10	1	Action n/g	.30
01.24.00	"	"	2	Cam wobble	.30
01.24.40	"	"	3	Cable in shot	.30
01.25.15	"	"	4	OK	.30
01.25.55	87-100	1.15	1	NG action	4.10
01.30.20	"	"	2	OK	4.05
01.34.30	96-100	"	1	Insert shots - poss	.45
01.35.25	"	"	2	" OK	.45

Example of typed up recording log.

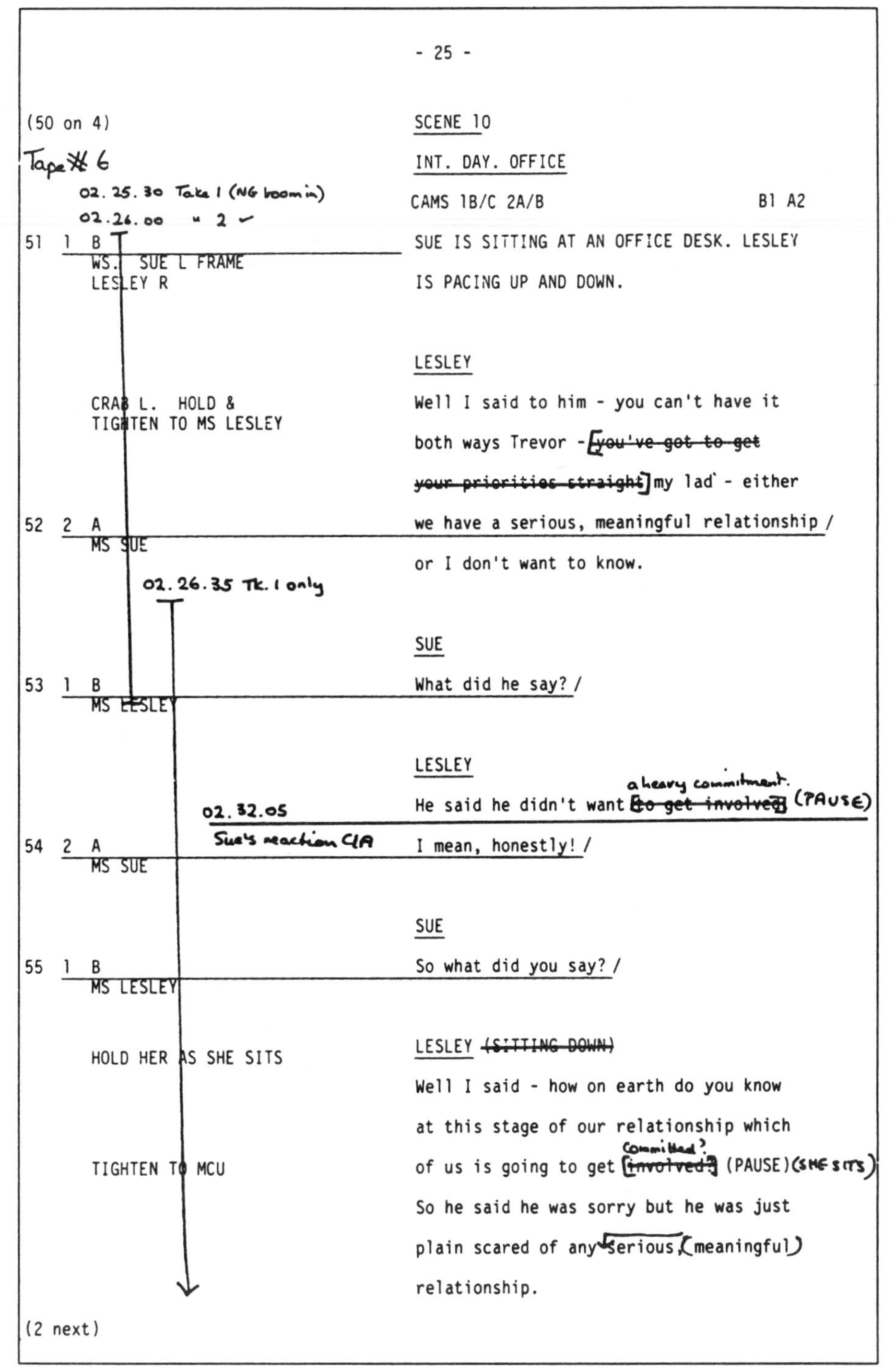

- 25 -

(50 on 4) SCENE 10

Tape # 6 INT. DAY. OFFICE

02.25.30 Take 1 (NG boom in)
02.26.00 " 2 ✓

CAMS 1B/C 2A/B B1 A2

51 1 B
WS. SUE L FRAME
LESLEY R

SUE IS SITTING AT AN OFFICE DESK. LESLEY IS PACING UP AND DOWN.

LESLEY

CRAB L. HOLD &
TIGHTEN TO MS LESLEY

Well I said to him - you can't have it both ways Trevor - ~~[you've got to get your priorities straight]~~ my lad - either

52 2 A
MS SUE

we have a serious, meaningful relationship /
or I don't want to know.

02.26.35 Tk. 1 only

SUE

53 1 B
MS LESLEY

What did he say? /

LESLEY

He said he didn't want ~~[to get involved]~~ a heavy commitment. (PAUSE)

02.32.05
Sue's reaction C/A

54 2 A
MS SUE

I mean, honestly! /

SUE

55 1 B
MS LESLEY

So what did you say? /

HOLD HER AS SHE SITS

LESLEY ~~(SITTING DOWN)~~

Well I said - how on earth do you know at this stage of our relationship which

TIGHTEN TO MCU

of us is going to get ~~[involved?]~~ committed? (PAUSE)(SHE SITS)
So he said he was sorry but he was just plain scared of any serious, (meaningful) relationship.

(2 next)

Example of coverage script.

Running time

The running time, taken from the stopwatch, should be written down.

These lists can be typed out after the studio, especially if the director requires a copy for viewing the recorded material before going into editing. If the director does not require a copy and if your notes are legible enough for you to read then don't bother to type them.

Coverage script

By compiling the recording log, you will have noted down exactly what happened during the studio in the order in which it was recorded. If you then attend the editing with this log and work from it by informing the editor of the correct rolls and time code points in which to re-assemble the material in its final story order, you might find it difficult and time-wasting as you hunt through your lists of notes in order to find the desired shots in the correct order.

This is where the coverage (or tramline) script is so valuable. The script, which you would mark up in rough on your own camera script during the studio, has the 'in' point of time code and lines down the page covering the length of the recorded sections.

If you have time during the studio, if not afterwards, mark up a clean camera script for the editing.

In this way you will be able to see *in story order* what is required for editing and you will speed up the whole editing operation.

5 Light entertainment and music

Light entertainment

Light entertainment covers a very wide range of programmes: situation comedies, stand-up comedians, quiz shows, music shows, chat shows and so on. The programmes are frequently studio-based but they might be made as an outside broadcast or on film or contain many elements within the same programme. They are often transmitted live or recorded 'as live' in front of an audience. The PA on light entertainment programmes will find herself using the whole range of her technical skills as she moves from one show to another.

Many of the skills she needs will be discussed in other sections of this book, but there are, however, a few points that specifically relate to these programmes.

Fluidity of style

Light entertainment programmes, even when pre-recorded, often tend to be more fluid than drama, more like live programmes in that there are often last minute changes. Many shows work from a running order rather than a detailed script.

Involvement of artists

Because many light entertainment shows revolve around the personality of one or more artists, those artists will tend to be far more involved in the setting up of the production than on other types of programme. The host of a chat show will probably attend planning meetings and have a certain say in the guests chosen to appear on the show. If a programme is centred around a specific comedian, the artist will expect to approve the script and dialogue and jokes may have to be altered to suit his or her personality.

Scripted shows

If there is a script it might well originate from a number of different writers. This can cause problems to the PA in the initial assembling of the material and she should bear in mind when it comes to the camera script that the document will be a rough guide only as there will doubtless be many changes made on the day of recording or transmission.

Estimating overall duration

It can be difficult trying to work out in advance the overall duration of the programme. Because pacing is the key to many comedians' routines, it is often not possible for the PA to work to the three words to a second rule which tends to hold good for presenters. Sometimes there might be pauses, for effect or for 'business' and comedians tend to speak at different speeds depending upon their style. Many of them ad lib during the recording or transmission and the PA will have to make allowances for all these factors when trying to work out timings. She will therefore have to work out whether to time hard on dialogue or adopt a more flexible system, bearing in mind that it will be difficult to be absolutely accurate on any assessment.

She will also have to allow for audience reaction time or time for 'canned' laughter where there is no studio audience and she must allow for applause.

In estimating the overall duration therefore, experience counts for a good deal. If the PA is familiar with the show, its style, its format and artists, she will be able to give a far more accurate assessment than one coming new to the programme.

'She will also have to allow for audience reaction time or time for "canned" laughter . . .'

Timings during recording or transmission

As there are often studio audiences for light entertainment shows, even programmes that are recorded are usually done 'as live'. Because of this and because many light entertainment programmes are transmitted live, the PA must consider timing to be one of her priorities.

If the programme is recorded she should take timings at virtually every line as she can then work out with the director the precise timing for essential edits.

Music

Music plays a very large part in many light entertainment shows. The timing of music is very important for reasons of copyright and possible editing and durations should be noted at every verse or phrase of music.

Bar counting

Another element in many light entertainment programmes is that of counting through music items, known as bar counting. It must be remembered, however, that bar counting is not, of course, confined exclusively to light entertainment.

Bar counting is an extension of shot calling in that the PA counts the bars of music in order to keep everyone in the studio aware of the point reached in the musical item. It is especially important in music which is solely instrumental. A camera operator might have to make a move after the eighth bar and would be listening carefully to the PA's count.

Bar counting is part of the PA's job which is sometimes viewed with alarm by otherwise experienced PAs who may have had little or no practice at the art and who might not be especially musical.

How to accustom yourself to counting bars of music

If you find the whole concept strange and difficult to understand you might find it helpful to approach the job in various simple stages. The first stage is to become aware of the rhythmic beat in different pieces of music. Some music has a stronger and more obvious beat than others. Listen to various types of music and trying tapping out the rhythm.

The next stage is to watch musical items on television and tap out the beat. Once you have the rhythm established you should try counting through the bars in each shot. With every shot change start again from one. This will begin to give you the feel of bar counting and will stand you in good stead when you come to do the real thing.

Music in your programme

If at all possible you should obtain a tape or CD of the music, take it home with you and play it over and over. Practise counting through the bars in the privacy of your own home and ensure that you can tap out the rhythm with confidence. The more familiar you are with the music, the better your bar counting will be in the studio.

If you cannot obtain the music in advance, don't panic.

Band calls

It is not likely that you would have to bar count 'cold' – that is without ever having heard the music played at least once. Most probably you would have a band call at which all the music would be played through. Good musicians will give you a bar breakdown at this stage. During the band call the director might work out his shots. He might then script them himself or call them out for you to note down. If the piece is instrumental you will need to count the bars between the shots as they are called out, in order to script them accurately.

'As directed'

If the sequence is 'as directed' you will have to rely on the bar breakdown given during the band call and the knowledge you have gained during rehearsals. Time each shot in the 'as directed' sequence.

What do you say and how do you say it?

You call the shot number and then count down through the bars. You will need to say 'Shot 51' rather than just '51' in order to differentiate between the shots and the bars.

So if shot 51 has eight bars you would say 'Shot 51 . . . 8 . . . 7 . . . 6 . . . 5 . . . 4 . . . 3 . . . 2 . . . 1'

If there is the possibility of losing the beat during the time you say 'Shot 51' then you should drop off the first bar, i.e. 'Shot 51 . . . (slight pause) . . . 7 . . . 6 . . . 5 . . . 4 . . . 3 . . . 2 . . . 1'

Depending upon the music you might or might not have time to call out the number of the camera you are coming to next. If anything has to be omitted, that should. If the shots come thick and fast it might be more helpful to the studio if you just called the shot numbers while retaining the beat of the music by tapping it out.

On complex music items or productions there might be two PAs in the gallery, one to bar count and the other to call the shots.

51	1 W/A ORCHESTRA. TRACK IN TO SINGER	(8 bars instrumental)
52	2 MCU SINGER	Snow fell in the city
		Snow fell in the town
	(8 bars)	Whitest drifts along the rivers
53	3 2-s BACKING	Echoes the blank in my heart./
		Da..da..da..da..do..do..do..
	(6 bars)	
54	2 MS SINGER.	Echoes the blank in my heart./
	TIGHTEN TO MCU	Rain falls on the city
		Rain falls on the town
	(8 bars)	It's rainin', rainin' in my heart
55	1 MS ORCHESTRA.	Since you let me down./
	TRACK R TO BACKING	Do..do..do..do..dum de dum..
56	2 (6 bars) MCU SINGER	Since you let me down./
		It's long, long since you left me
	(4 bars)	
57	1 MS ORCHESTRA	It's long, long since you're gone/
		(4 bars)
58	2 MCU SINGER	But the cold remains
	(6 bars)	The hurt, the pain,
58	3 2-s BACKING	The emptiness in my heart./
	TRACK BACK TO 3-s with SINGER	The emptiness in my heart.
	(8 bars)	Do do de da do do..do do de da do..
		Do do de da do do..do do de da do..

Example of typed and marked up script of song.

Beating time

It helps to beat time when bar counting. Use a ruler, a pencil, anything, but do retain that firm, strong beat. There might be a metronome in the gallery to help you.

Speak loudly

When bar counting it is especially important to speak clearly and loudly as your voice will be competing with the music.

Losing your place

If you lose your place or get it wrong when bar counting, my best advice is to keep quiet until you are sure you can restart correctly. That might be at the beginning of the next shot or on a fresh line of the song. What ever you do, do not try to catch up as you will only confuse everyone. Just keep the rhythm going by tapping it out, preview the shots and start again when you feel confident.

6 Live and as-live programmes

Live and as-live programmes might vary in content and complexity, but they do all have one thing in common. When the PA is sitting next to the director in the studio gallery or outside broadcast control room, counting down to transmission, outwardly calm and in control, her heart will nonetheless be pounding away, her mouth dry, the palms of her hands will be damp and adrenalin will be pumping round her system at overtime rate.

What is also certain is that once on air these symptoms will cease. And it has been truly said that if a PA stops having these symptoms, it is time she gave up

'Outwardly calm and in control . . .'

working on live shows. Live programmes are hard, exacting and demanding, but they do have their own excitement.

The demands of the job

The role of the Production Assistant in the gallery of a programme that is being transmitted live or 'as-live' is one that is often technically very demanding. It carries with it a weight of responsibility and requires coolness, concentration and the ability to work under extremes of pressure. An aptitude for fast, accurate mental arithmetic is a definite advantage.

Responsibilities

The job can be summed up in general terms by stating that a PA's function on a live show is to get the programme off the air cleanly and on time, but to say that is an over-simplification, for the PA is responsible for a good deal more besides, namely:

- taking overall and insert timings in order that the producer may be able to allocate the requisite time for each individual item within the total duration of the programme,
- being the central point of communication both for receiving and disseminating information,

- liaising with network control to get on and off the air smoothly,
- taking in changes while on the air and adjusting timings accordingly,
- keeping everyone connected with the programme informed as to the point reached on the running order or script by means of verbal idents and shot calling,
- giving warning standbys to those who require it, e.g. presenters, sound, VTR operators and so on,
- providing a smooth transition from live to pre-recorded material and back by means of accurate time counts,
- acting as a verbal clock, giving countdowns to presenters and timings to the end of each individual item and to the end of the programme,
- bar counting if musical items are involved.

The Production Assistant may be required to do all or only some of these things and possibly has other responsibilities during the gallery which I have not listed, depending on the nature and content of the programme. Many of the same jobs are also required of the PA on pre-recorded programmes, but if the show is live, there is the heightened tension, the knowledge that mistakes cannot be edited afterwards but will be seen by the entire viewing public. This thought makes many experienced PAs shy away from the thought of doing a live show, but it also gives the extra edge, the added excitement and the attraction for other PAs which they find lacking in pre-recorded programmes. The fact that once the programme is over it is totally finished with is another bonus. Mistakes are soon forgotten and there is little clearing up and the minimum of paperwork.

So, having arrived in the gallery, what are the procedures?

Script check

The director may well take the studio through a script check, running rapidly over each link and item, talking to the camera operators about what is required and ironing out any problems.

Rehearsal

If there is time, there may well be a rehearsal of one or two potentially difficult bits, for example, links into items that need precise timing, complex shots using electronic effects etc. These rehearsals would give the PA the chance to adjust roll cues if necessary.

PA's role

During this run up to transmission, the PA will be fully occupied with a variety of different jobs. In no particular order these could include:

- receiving further pages of script, putting them in sequence order, marking up the fresh script pages with roll cues and standbys,
- on receiving the edited durations of the insert items, the PA will note down the accurate timings on her chart, adjust the cumulative times accordingly and keep up to date with the overall duration of the programme,
- she will liaise with the producer on the question of overall duration, noting any changes and informing the director and studio,
- she will check that her talkback is operating, make contact with the operators of the VTR channels and run through the order of items with them, checking the identity of the source channel for each item,
- she will check that her stopwatches are working properly and find out from the studio supervisor whether the studio clock is running to correct time,
- she will make contact with presentation and liaise concerning the times in and out of the programme together with times for commercial breaks should these occur,
- she will make contact with other studio centres or outside broadcast units for live injects into the programme,
- if the overall time as given by presentation differs from the original running time, the PA will adjust her timings accordingly and inform the producer and director,
- the PA will check the order for capgens,
- if she has a digital countdown watch she will pre-set it to the exact running time of the programme,
- if the telephone rings either the PA or the studio supervisor will answer it.

In addition to all the above the PA *must* at all times, keep an ear open for the director, count in and out of items being rehearsed and make changes to the script and time chart where necessary.

Recording off transmission

It is the policy of some companies to make a broadcast quality video recording of a live transmission for archival or other purposes. If this is not done, the PA should ensure that a non-broadcast quality recording – probably VHS – is made of the programme as it goes out. This is invaluable for possible legal purposes should any part of the programme content be questioned. It is also very useful for the PA when timing music, archive film etc. for copyright purposes.

Five minutes to transmission

The PA should give minute warnings to the studio from about five minutes prior to transmission. Certainly a one minute warning must be given followed by a thirty second warning, then fifteen second, then counting down from ten.

Standbys should be given for the channels needed for the opening sequence, the digital countdown clock started and the programme counted on the air from −10″.

The opening titles should be rolled and at zero the overall and insert watches started.

'During this run up to transmission, the PA will be fully occupied with a variety of different jobs.'

Different systems

There are a number of different systems in use for running a live programme, dependent upon custom and the complexity of the programme.

In a straight news programme, with its great number of items, its liability for very last minute changes while on air, its need for absolute split-second precision in timing, it may be that the PA is unable to do anything during the transmission other than overall timing and countdowns. Then the director will give the standbys and roll and cue the sources.

On some live programmes the complexity of the content will require two PAs to be in the gallery, one to time and give countdowns, the other to roll and cue.

But many live programmes rely on just one PA who is responsible for all these elements. Very often, in this situation, the director does his own vision mixing. We will assume, in this section of the book, that there is just one PA in the gallery.

A question of priorities

If we break down the PA's job in a live gallery into its component parts we find an order begins to emerge. Some things are of greater importance than others.

Timing

It is of over-riding importance to get on and off the air smoothly and cleanly. Everything else is subservient to that first priority although from it naturally flow things like good liaison with the producer and rolling into and counting out of insert items correctly.

Previewing

All PAs, whatever type of programme they work on, must watch the preview monitor for the shot that is to be transmitted next.

Rolling and counting

Having ensured that the beginning and end of the programme are correct, the PA's next priority is to make the body of the programme as smooth as possible by rolling in the insert items at the correct times and counting in and out of items.

Standbys and cueing

It is important to give warning standbys, but it does not head your list of priorities, neither does cueing in capgens. If necessary, they must be dropped in favour of more vital work.

7 Learn to love your stopwatch

A PA without a stopwatch is like a fish out of water. Your watch is part of you, you should feel naked without it. It is the core and the essence of the live PA's job.

To know your stopwatch requires far more than the mere ability to stop it, start it and flick it back to zero. You have to be able to read it, not just in the sense of being able to tell the time quickly and accurately, but to be able to work out instant calculations from the information it gives.

Some PAs find it helpful to think of the face of the stopwatch as an orange which they divide into segments: first quarter, second quarter and so on. It might sound fanciful but if you are giving countdowns through a pre-recorded item you should be able to glance at your watch and know that in an overall duration of 55″, when the second had has reached 25″ there will be 30″ left on the insert, and if thinking of the face of the watch as an orange or apple or football helps, then you should think of it as such.

Analogue and digital

This is where a stopwatch with an analogue face has the advantage over a digital watch. A digital watch will only show you the time it is now, whereas with an analogue watch you see the time in relation to what is past and what is to come.

Therefore in giving countdowns out of an insert you are not just keeping a lot of abstract numbers in your head, you are looking at the 30″, the 15″ and so on as a shape, a shape on the round face of the dial from the position of the second hand and in relation to the overall duration of the insert. Back to the orange in fact!

That is not to say that time shown in digits is of no use. It is, especially in a digital countdown clock which can be pre-set with the programme's running time.

Digital countdown clock

If the running time is 18.32″ then you can pre-set those figures and, by starting the clock as you go on air it will count down to zero.

As you will be extraordinarily busy with other things at the precise moment you go on air you might like to pre-set your countdown clock a minute or two early – setting it for the programme running time plus one or two.

The advantage of a digital countdown clock is that at any point you, or anyone else can see exactly how long remains to the end of the programme. It is a useful back-up in situations where things have gone wrong and you no longer have any idea of the overall timing, and you can use it to advantage to get you off the air, but do not rely on it exclusively, firstly because it can and sometimes does go wrong and secondly because only certain companies provide this facility.

These clocks can also count up from zero or from any preset time and therefore can be used as normal stopwatches.

Number of stopwatches

It might have seemed rather excessive when I stated that you should take a number of stopwatches to the gallery but watches have been known to break down and to PAs whose master overall watch has unaccountably stopped in the middle of a live transmission – and it has happened to a surprising number of PAs – a back-up master then becomes essential.

At the beginning of the programme you start your one or two master overall watches and put them some way out of reach so that you are not tempted by their proximity to confuse them with an insert watch and zero them by mistake – and that has been known to happen before now!

If you feel that you might in any way be confused by the plethora of stopwatches spread before you, you could always use a digital watch for overall timings and an analogue watch for inserts, or you could mark the glass of the master watch with a red pen.

Using just one watch

On some news programmes the PAs work solely from just one stopwatch. They do this because it is less time-wasting than going from one watch to another and because there is less risk of confusion.

They work off one watch by noting down on their running order the overall time *as they go into* a pre-recorded item. They then add to this figure the duration of the item and the cumulative figure they reach gives them the basis for working out countdowns to the end of the item.

For example, if the time on going into the item is 12.30″ on the overall watch and the duration of the item is 1.20″, then the 'out' time of the item will be 13.50″. You can then give countdowns based on that time: 30″ countdown at 13.20″; 15″ countdown at 13.35″ and so on.

This system has the advantage of being able to give accurate overall timings at the *start* of each item, rather than at the end. It means that you have only one watch to worry about, rather than several but it is not easy to master the technique and does rely on a lot of timings being written down which presupposes that the PA is solely concerned with timings and does not roll and cue as well.

Real time versus stopwatch time

PAs on some programmes work not from a master overall stopwatch, building up the running time of the programme from zero, but from the real time as

Learn to love your stopwatch.

shown by the studio or digital clock in front of the PA. They argue, very validly, that because the programme is live, it is the real time that matters, i.e. getting off the air at 18.32.35.

On their running order or time chart therefore, these PAs will add up their cumulative times in the context of the real time and backtime from the real off air time.

The advantage of this system is that the studio clock and their own timings match and their backtimings for getting off the air will be correspondingly simpler.

The disadvantages to this system are:

- if the running time of the programme is always the same but the on air time varies and is not known until just prior to transmission, the PA will not have been able to work out backtimings in advance,
- if the running time of the programme is subject to last minute changes the PA will have to re-do quite complex sums as it is harder to work in large figures than in small. Sums involving six digits seem more daunting than those with four.

If the PA is working off real time, however, it is essential that she either has a digital clock in front of her on the desk or the studio clock is clearly positioned for her use.

Counting down from 10″

But having so far stressed the importance of learning to read your stopwatch at a glance, you should make it your business to learn how to count down from 10″ without reference to one. It will make you less dependent upon your watch and more able to preview the monitors at critical moments.

8 Rolling, cueing, counting

Programmes that contain a mixture of source material are broken down into sections, often called sequences. The PA has to make sure that:

- there is a smooth transition from one sequence to another, i.e. from live to pre-recorded and back, and
- that everyone knows what is going on at every stage of the recording or transmission.

This involves overall timing, which is covered in the next chapter, counting through items and cueing various actions to take place at the correct time, for example rolling in VTR or pre-fading music.

Rolling insert items

The instruction to the operator to start the machine containing the pre-recorded insert is sometimes given by the director, but more usually by the PA.

Standby warning

First of all a standby warning is given at anything from thirty seconds to a minute before the item is to be transmitted. The videotape operators will respond by a 'beep' to show that he or she has taken heed of the warning. If there is no response to the standby, the PA *must* repeat it with more urgency.

In some studios the PA herself operates the VTR machines in which case it is obviously not necessary for standby warnings to be given, apart from mentally!

Ident clock

The ident clock should appear on one of the monitors. The PA should check that the correct story title is shown.

Run-up time

Each machine needs a certain time in order to run up to speed. These times vary and the PA will have worked out, on the basis of the run-up time needed,

'If there is no response to the standby, the PA must repeat it with more urgency.'

when to give the command for the machine to be started. This can be worked out:

- On a word cue. Counting back three words to a second from the end of the presenter's link into the item will give the precise word on which the PA gives the command to the operator. This cue can be adjusted in rehearsal if the presenter speaks faster or slower.
- On a time cue. If the presenter's link does not contain sufficient words to allow a word cue, or if the inserts are rolled back to back (one insert leading straight into another insert), or if there are inserts within the pre-recorded item, then the PA has to work out time cues for the command.

- On an action cue, e.g. lighting a cigarette, sitting down, or whatever has been worked out in advance between the presenter and PA. This would occur in situations where there is no precise timing in the live section before the pre-recorded insert and no scripted link.

It is becoming increasingly usual to use a so-called instant start, which means that in theory the PA does not have to give a pre-roll cue. In practice one should still allow for the time it takes for the operator to respond to the command, say one second.

Command

Different companies use different command words and the order in which they use them varies, i.e. 'Roll VTR', 'Run VTR', 'VTR Roll', 'VTR Run'. Everyone, understandably, defends the system they are used to and clearly one must give the recognized and explicit form of command that is common at the time.

Cueing

The director often cues whatever action is to take place. All the PA does is to give standby warnings.

Presenters

However, the PA usually cues the presenters when they have live out-of-vision dialogue to speak over a pre-recorded insert.

Capgens

The PA is often required to cue the capgens, counting down to the name supers.

Cue dots

These are visual warning marks, in the form of small squares which appear in one corner of the television screen and disappear during the last minute or thirty seconds of a programme or leading to an opt or commercial break.

Cue dots are for the use of presentation or the PA in a live gallery, who will be watching for the cue dot of the preceding programme in order to count her own programme on the air.

In the UK, the BBC place the cue dot at the top left hand side of the screen. The cue dot appears at 30″ before a live transmission and disappears at 10″ before transmission. The ITV companies place the cue dot at the top right hand side of the screen. It appears at 1′ before a junction or commercial break and disappears at 5″ before the junction or break. It is sometimes the PA's responsibility to operate the cue dot.

Just keep counting

Counting is a major function of the PA throughout a live transmission. She will:

- count down into the programme,
- count into, through and out of every pre-recorded item,
- count capgens onto the air,
- count presenters into live voice-overs, into, through and out of interviews,
- count out of the programme.

Counting into the programme

The PA should keep the studio informed at regular intervals before they go on air. It is usual to give at least a five minute warning, then warnings every minute. From one minute to transmission she will give a 30″ count, a 15″ count and count down from 10″ into the programme.

Inserts

The PA should count into the pre-recorded inserts, counting down the number of seconds remaining on the VTR clock.

Once into the pre-recorded item, the PA should give the total duration of the insert, i.e. 'Three minutes on VTR, three minutes', then warnings every minute, 'Two minutes left on VTR, two minutes'. From one minute she should give a 30″ count, then 15″ and count out of the insert from 10″.

Capgens

If the PA knows that there is a capgen at 1.30″ into the insert, she should either give a 5″ countdown to the superimposition and, if required, count through the 5″ or so that the director wants it held on the screen. Alternatively she could count the capgen on air by saying 'capgen at 1.30″ – 27, 28, 29, 30'.

Commercial breaks

These are treated by the PA as if they are insert items. Presentation will inform the gallery how long the break is to be and the PA will count to the break, start an insert watch for the duration of the break and count back to the studio at the end.

Counting out of programme

The count out of the programme is usually complicated and requires a list or box of timings worked out in advance to enable the PA to get off the air smoothly. She would certainly give a 'one minute to end of programme' warning and then a 30″ warning, 15″ warning and count out from 10″. Ending the programme is covered in Part Three chapter ten.

Is anyone listening out there?

It is possible to feel very isolated in the gallery, despite being surrounded by people and knowing that when you speak you are communicating with many others. It is possible for you to sit at the control desk, counting in and counting out, and a small corner of your brain can be wondering, just what is the point

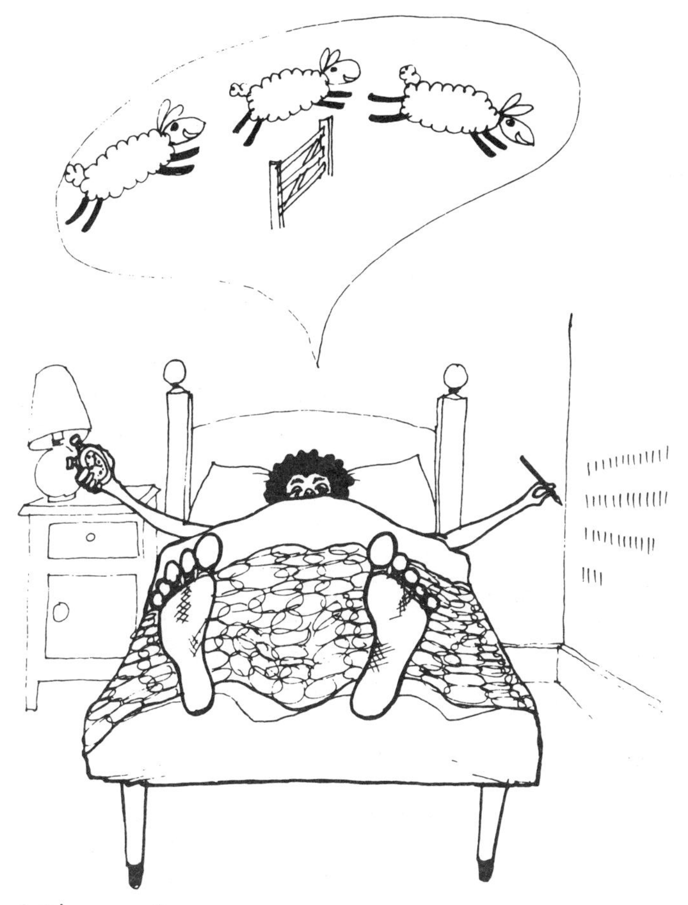

Just keep counting.

of it all? Is anyone listening out there? If you stopped counting, surely the programme would continue, unchecked, except for a blessed silence?

Yes, it most probably would continue, but you would throw everyone into a good deal of confusion. For they rely on the PA to guide them through the programme, to keep them informed about what is happening at that moment, what is to happen next, and how long they have, in terms of time, at every stage of the programme.

9 Studio discussions and interviews

Studio-based discussions and interviews take up an increasingly large amount of television. Such programmes are usually transmitted live or recorded 'as live' with the minimum of post-production work.

An interview can range from a simple injection into an otherwise news orientated programme when a political figure or 'expert in the field' is asked to comment on a news item, to a whole programme structured around a person-to-person talk of the 'chat show' variety. Then there are programmes involving discussions between groups of people. These frequently include the studio audience.

Whatever the situation, because these discussions or interviews are more or less unscripted, the television coverage will have to be 'as directed'. In other words the direction of the shots will be ad libbed by the director as the programme develops.

Rehearsal

There might be some form of rehearsal for the show, but it would tend to be minimal as anything too well-rehearsed and detailed would destroy the freshness and spontaneity of the programme. The most that would happen in advance would be the seating of the participants, the allocation of cameras and the working out in general of the size of shot each camera would undertake. Perhaps some of the questions or areas of discussion might be touched upon with the interviewee(s).

Counting down

The PA's first priority in the gallery will be to keep a keen eye on the time and count through the interview or discussion. She will know in advance how much time has been allocated to the item and she *must* keep everyone informed as to the length of time remaining.

Her counts will vary according to the length of the interview or discussion. If it is a short interview of only a minute or so duration she will have to give precise and frequent countdowns. But if the entire programme – say of 45.00″ duration – is a discussion then she should only give countdowns every 15.00″ until the final 15.00″. From then she should give warnings every 5.00″ and count down in the usual way from 1.00″ to the end of the programme.

Presenters

Never give a presenter the exact minutes and seconds they have available for an interview or discussion, i.e. 'You have four minutes thirteen seconds on this interview', as it is confusing. Always round up or down to the minute or half minute, i.e. 'You have just over four minutes on this interview.'

During an interview, the PA should give the presenter counts to the end, either directly or via the floor manager. Some presenters like to be counted out of interviews and some do not. The PA should find out in advance what system the presenter prefers.

As directed

Any 'as directed' sequence should be clearly stated as such when the PA is in the gallery and she should also name the cameras involved. From then on it might be helpful for her to inform the studio what is happening at each moment, i.e. 'on camera 3'.

Assisting the director

In certain circumstances the PA could greatly assist the director by listening to the discussion and helping assess what is likely to happen next. That sounds very much simpler than in fact it is because most PAs find that they emerge from a studio gallery with little or no knowledge of the content of the programme with which they have just been involved. This is especially true of the more technically complex programmes.

But in the studio discussion/interview situation this kind of assistance would be very helpful to the director.

Say, for example, the PA was involved in a programme comprising a panel of people answering questions from a studio audience. The audience would have been specifically selected with groups representing specific interests. There would be a seating plan and the PA would know in advance where each group was seated and which camera was closest. By listening to the discussion and by constant reference to the seating plan she could advise the director, camera and boom operators should a shot be needed of a particular person or group.

Recorded programmes

If the programme is to be recorded with an editing session to follow, the PA should make notes for the editor, i.e. the questions asked, comments of particular interest, anything which definitely needs to be edited out of the final programme and so on.

'In certain circumstances the PA could greatly assist the director by listening to the discussion and helping him assess what is likely to happen next.'

Linking to next item

If the presenter is to close the interview/discussion with a scripted link from which the PA is to roll a pre-recorded item then she should be alert to any possible re-wording that the presenter might have to make in order to effect a smooth transition from the preceding item. If the link is unscripted then the onus is on the presenter to ad lib for the necessary time after the PA has rolled the item until it appears on the screen.

10 Overall timing

How does the PA keep a running total of the overall timings amid the welter of rolling, cueing, counting in and out and all the other sundry duties undertaken while on air?

This is how it is done. At the end of each sequence you should note down the cumulative time which you will get from your master stopwatch – unless you are using just one watch as mentioned in the last chapter. You note this down either in the 'on air' column on your time chart, or on your script if you work solely from that.

By comparing this figure with your estimated cumulative timing you will be able to work out, at the end of each sequence, whether the programme is running to time, under or over. You inform the producer and he or she will then make the necessary adjustments to enable the programme to come out on time.

Producer's options

There are a number of options open to the producer:

- the buffer item, i.e. the interview, could be lengthened or shortened,
- if the programme is under-running, the presenter could be asked to ad lib for a few seconds, or a standby item could be inserted,
- if the programme is over-length then some lines could be cut from the presenter or a sequence dropped altogether,
- presentation could be contacted and permission sought for an over-run of the allotted time.

What is essential, however, is that the producer is given the information by the PA, and that it is given with enough time for the different options to be considered and carried out.

When do you note down the timing?

If, at the end of each sequence, you find you are too busy to glance at your overall watch and write down the result, you might find that you get into a system of glancing at your master watch at, say, one minute to the end of the insert when you have slightly more leisure. You could write down the figure, adding on one minute at that point, thus leaving you clear to count out of the item.

Ending the programme

All these timings are of course carrying you inexorably closer to the end of the programme and as it progresses you will know by how much or how little the programme running time has changed. If you look at the time chart below you will see that by the time we reach the end of Sequence 7 we are 10″ over-running. This is not serious and the producer will probably make the decision to cut 10″ from Sequence 12, the closing headlines.

We could then work out the end of the programme precisely by backtiming from the 'off air' time (see the far column on the time chart). Whatever happens, at 20.10″ we *must* go into the closing titles which are of 15″ duration and recorded on VTR. That means that at 20.05″ *precisely* (allowing for the 5″ run up time for the VTR machine) we must roll VTR.

	Est'd	Cum	Actual	Backtiming	On air
1. IDENT/TITLES	0.30	0.30	0.30		0.30
2. LOST CHILDREN	2.35	3.05	2.05		2.35
3. NEW TECHNOLOGY	2.20	5.25	2.05		4.40
4. SOCCER VIOLENCE	1.15	6.40	1.45		6.25
5. SHORTS	2.45	9.25	2.15		8.40
6. TEASE	0.15	9.40	0.15		8.55
7. WAR CRIMES	2.10	11.50	2.05	10.50	11.00
8. STATE VISIT	1.25	13.15	20.05	12.55	
9. HIDDEN TREASURE	1.55	15.10	1.55	14.50	
10. SPORT	2.35	17.45	2.50	17.40	
11. WEATHER	1.15	19.00	1.20	19.00	
12. CLOSING	0.35	19.35	0.35	19.35	
13. FINALLY/VTR	0.50	20.25	0.50	20.25	

Example of a time chart.

Final minute

As there will be a lot happening at the end of the programme, *it is essential* to write down a list of the final timings from the 1.00″ to end warning.

This last minute of the programme should be counted through either from the studio clock, if you are transmitting live, or from your master stopwatch if you are recording the programme 'as-live'.

On the final page of the script we would work out a box of final timings which could look something like the example below:

19.25″	1.00″ to off air/S/B VTR
19.55′	'15″ left' warning to presenter/30″ to off air
20.00″	'10″ left' warning to presenter/S/B roller
20.05″	'Roll VTR' (perhaps linked to a word cue, perhaps rolled on time alone if the presenter is ad libbing the closing words)
20.10″	'On VTR for 15″'
20.15″	10″ countdown to off air
20.25″	Off air

Coming to the end of a live transmission depends very much on the composition of the last couple of sequences of the programme, but the countdown timings to coming off air are, for safety's sake, better worked out and written down as there might be a lot of things to say and timings to give and it is vital that nothing is left out.

One minute clock

Because the last minute of live transmission is crucial, some presenters like to have a clock started in the studio at one minute to the end of the programme in order for them to time their closing links precisely.

This clock might be set by the PA from the gallery or by the floor manager in the studio on instructions from the PA.

Closing credits

There might well be closing credits that will have to be cued in from the capgen machine at fixed times. These times would be added to the box.

Pre-fading

Music

If there is closing music, it is far nicer for the programme to end with the end of the music. This means that the music (on tape, CD or disc) must be started before the sound supervisor fades it up. This is called 'pre-fading' and usually happens on a cue from the PA.

SEQUENCE 13

/S/B VTR/ | /S/B PREFADE/

CAM 1 /
CHERRY IN VISION

CHERRY

And finally, a happy story from Loudwater Zoo. Two weeks ago, Henry the elephant lay down in his cage, spurned all food and grew progressively weaker. After trying every possible remedy, John Tombs, his keeper, finally advertised for a mate for Henry. Now, two days after the arrival of Nelly, Henry has taken on a new lease of life. Goodnight. /

Q PREFADE GRAMS AT 19.45

ROLL VTR-1

CAM 2
2-s CHERRY AND ANDREW

VTR-1 / 19.55
HENRY & NELLY AT ZOO

/F/UP GRAMS/

CLOSING CREDITS

DUR: 40" available

CAPGENS

Presented by
Cherry Stone and Andrew Plum

Producer

30"

20.25

Example of pre-fading music. On a script, sound details are always written on the far righthand side of the page. During rehearsal, confirm with the Grams Operator the length of the closing music, and ask him to set up the grams for your pre-fade. On many live programmes the Grams Op will pre-fade the closing music on his own timings. It is therefore most important to tell him of any alteration to the running time.

Let us say for example that the quarter-inch tape containing the music runs for an overall duration of 40″, while the duration of the closing credits is 30″. We therefore have 10″ of music which can be faded in under the closing words of the programme.

Clearly the end of the music must coincide with the end of the programme therefore the PA would calculate 40″ back from the programme end. If the programme was 10 minutes long, then the music would have to be pre-faded at 9.20″. At 9.30″ the cue would be given to 'fade-up' the music. The music would then end with the end of the programme.

VTR

You might also have to pre-fade a VTR insert if the programme end was to coincide with the end of the insert.

11 Hazards of the job

Just what can go wrong when working on live programmes? Well, almost anything.

Problems, which ultimately end up in the gallery, can arise from many different sources:

No time, no time!

For example:

- last minute script changes,
- late arrival of edited inserts,
- the running order being changed while on air,
- the total abandonment of the running order because a major story has just broken.

More often than not this lack of time, this last minute rush, is because the programme needs to be as topical as possible and this is especially true of straight news programmes. If, as a PA, you cannot cope with this situation you should be working in another more leisurely area of television.

If, however, the live programme is not of this nature but there is nevertheless a terrible last minute panic it could indicate perhaps an indecisive producer, journalists and researchers who have got into lazy habits or a general unprofessionalism which might be worth examining in order to see whether the situation could be improved.

Technical problems

These can be caused by human error, e.g. the wrong insert item appearing on the screen, the wrong captions punched up or the right captions appearing over the wrong picture . . . the possibilities are endless. Sometimes these mistakes could be avoided by the PA making sure that she previews the monitors, checking for the shot that is to come next.

Technical problems can also be caused by circumstances beyond anyone's control. Machinery sticks, jams, breaks down. Considering the things that could go wrong it is, in one sense, amazing that so many live programmes occur without any apparent hitches.

Your priority, as PA, must be to get back on sequence whatever happens, and worry about the timings later.

'Problems . . . can arise from many different sources . . .'

Your own special problems

Apart from coping, within the limits of your responsibility, with any of the hazards cited above, you might well have a few personal ones of your own:

Forgetting to wind your watches

This should be one of your first tasks in the gallery, or before you go into the gallery.

Forgetting to start your master stopwatch

If this happens you can always work from the main studio clock, or from your digital countdown clock if you have one. You could also start your master watch at the end of the first sequence as you would have a definite timing for that.

Forgetting to start your insert watch

You can work out the timing from your master watch, i.e. at the end of the preceding sequence you have noted down the overall time to be 9.40″. The link into the insert is 10″, making the time you went into the insert 9.50″. The time now, on your master watch is 10.00″, therefore you must be 10″ into the insert.

You could start an insert watch at this point or work off the overall watch, i.e. if you went into the insert at 9.50″ and the duration of the insert is 1.10″ then your 'out' timing must be 11.00″. You can count out to that.

But supposing you did not note down the overall time at the end of the preceding sequence? Perhaps you were fully occupied with some last minute changes to the running order or some other crisis at the time. What you *do* know, however, is that the overall length of this particular insert is 2.40″. You would have a rough idea that you had let, say, 10 to 15″ elapse before you (a) realized your error, (b) panicked, and (c) decided what to do. You could then give the studio an approximate idea of how long it was until the end of the item. 'About one minute left on VTR'. It is better than nothing. And do keep stressing the 'out' words.

An alternative might be to take your timing from the first capgen. Let us say that it is to be cued into the insert at 26″. You will, of course, be unable to cue it in but someone will do that, either the director, the producer or the reporter and you can then start your insert watch knowing that 26″ have elapsed since the start.

Forgetting to start your digital countdown clock

It is not a vital piece of equipment, provided you have other means of timing and use it as back-up only and an aid to getting off the air. If you forget to start it at the beginning of the programme, you could always start it late, providing you alter the countdown timing to coincide with the remaining length of the programme.

If your insert watch breaks down

You can work out where you are in the insert from your master watch using the method described above.

If your master stopwatch breaks down

You could work off the main clock. For example, if you know that the cumulative timing at the end of Sequence 11 was 8.35″ and your master watch has broken somewhere in the middle of Sequence 12, you know that your 'on air' time, was, say, 18.00.00 and that the time now, reading off the studio clock, is 18.09.35, you would therefore be 1.00″ into Sequence 12. You can then either work out the rest of your timings using real time or start another watch. Of course the examples I have given are simple and the timings will be far more complex than that, but if you grasp the system then you could put it into practice at need.

Alternatively you could work off your digital countdown clock, working out your backtimings from zero in order to match the time on the clock.

If the studio clock is inaccurate

You should always check, before the start of the programme, whether the studio clock is accurate. If it is a second or two fast or slow you could still work from it, adjusting your timings accurately.

If the studio clock breaks down (and it has happened!)

If all your fail-safe systems fail, don't forget that you can always ring presentation. They will give you the duration to the end of the programme and you could start a watch from that, or they might give you a timing, say at ten minutes to the end, to enable you at least to bring the programme out smoothly.

If you have no accurate timing for the insert

If you know the timing is inaccurate you should give the studio the approximate timing and stress the 'out' words.

If you have no accurate timing for anything

You might go into the gallery with few, if any, timings. In that situation remember that the end of the programme is more important than the beginning. As soon as you can, start backtiming and work out your last minute of the programme cues so that at least you come off air cleanly no matter what happens before.

If the running order is abandoned

It might well be changed on air for various reasons, in which case adjust your timings and remember the advice above. If it is abandoned altogether it will be because of special circumstances, perhaps a major news story that has broken. Presentation might well allow you to be virtually open-ended with no fixed 'off air' time. If not, you should backtime from the known off air time and give the studio firm countdowns. If no-one heeds them, it is not your fault.

If your sums are wrong

Always double check your figures for it is vital that they are accurate. If you are wrong and realize your error in time, there is no harm done, although your confidence will be shaken. If your sums are wrong and you don't realize it until

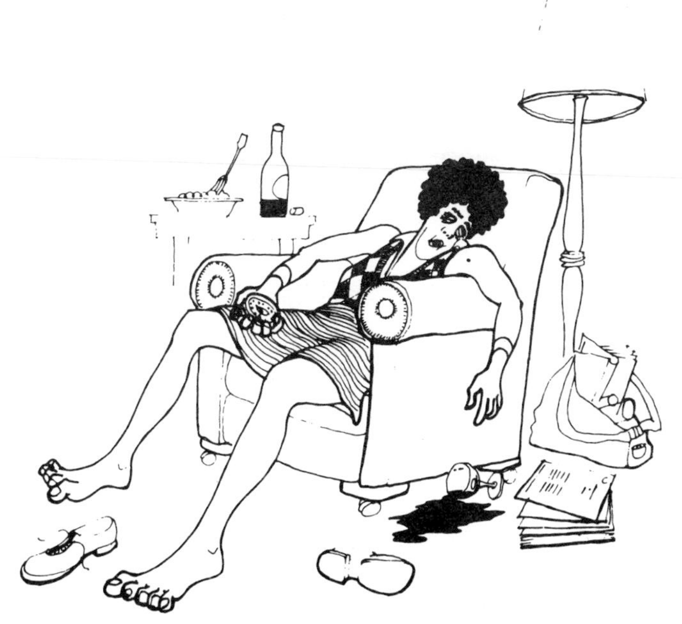

'... go to sleep and forget about it. Tomorrow's programme will go like a dream!'

too late when you have perhaps crashed off the air, or been taken off by presentation, your options are:

- to go out and shoot yourself,
- to hand in your resignation,
- to determine to do better next time (if you are allowed a next time)!

Don't forget that everyone is human and subject to error and that even PAs are allowed to get things wrong once.

And after this horrendous programme where everything that could go wrong has gone wrong, I suggest you go home, have a good stiff drink, go to sleep and forget about it. Tomorrow's programme will go like a dream!

12 Putting it all together

In the preceding chapters we have looked at the work of the PA in the studio gallery on live and as-live programmes, isolating in turn the different elements that go to make up her job.

It is difficult, in a book, to fit it all together in terms of exactly how a PA conducts herself during a live transmission. Difficult because the experience needs to be gone through if not for real, then at least in a practical exercise.

But in order to give a flavour of what a PA does and says, step by step through a programme being transmitted live, we will work through the following imaginary programme of ten minutes duration. It is a simple show, with just four sequences:

		Estimated durations
1.	Opening titles (VTR)	30″
2.	Link into HIGHLAND GATHERING (VTR)	3.50″
3.	Link into studio INTERVIEW	4.30″
4.	Closing link into VISION OF THE HIGHLANDS (VTR)	1.10″

For reasons of space I have typed the script in a continuous form, normally each sequence would be typed on a fresh page.

THE AFTERNOON PROGRAMME

/S/B VTR/

/ROLL VTR-2/

VTR-2 : OPENING TITLES	Duration: 30"
CAM 1 / MARY IN VISION S/B VTR-6 ROLL VTR6	MARY Good afternoon. In today's shortened version of The Afternoon Programme, we will be bringing you a report from the gathering of the McGuiness clan, followed by an exclusive interview with Hamish McGuiness, self-styled King of the Highlands. But to begin with, John Tew reports from what must be one of the strangest meetings in the western world.
VT-6 : HIGHLAND GATHERING	DURATION: 3.50 CAPGENS: Margaret McGuiness @ 1.05 Colin McGuinness @ 2.25 OUT WORDS: -- in all of Scotland.

/CAM 1 next/

CAM 1 /	MARY
MARY IN VISION	With us in the studio we have Hamish McGuiness, known to his
CAM 2	clan as King of the Highlands./
2-s MARY and HAMISH	Hamish, where does the title come from?
CAMS 2 & 3	AS DIRECTED INTERVIEW
CAPGEN: Hamish McGuiness DURATION: 4.30	
/S/B VTR/	/S/B GRAMS/
CAM 1 /	MARY
MARY	Well thank you very much indeed. And that's all for today. We'll be back at the same time tomorrow with a rather different view of our
/S/B CLOSING CREDITS/	national sport. Meanwhile
ROLL VTR	(we'll) leave you with some music and pictures of the misty Kingdom of the Highlands.
	/GO GRAMS/
VTR : VISIONS OF THE HIGHLANDS	CLOSING CREDITS
Dur. 55"	Presenter : Mary Calloway Director : Leonard Greene Producer : Philip Smith

Prog. No: ABC/123

RUNNING ORDER : THE AFTERNOON PROGRAMME

Wednesday 25 November 1992

In: 15.05.00
R/T: 10.00
Out: 15.15.00

Producer : Philip Smith
Director : Leonard Greene
P.A. : Louise Blair
F.M. : John Smith

Presenter : Mary Calloway

SEQ	SOURCE	TITLE	EST.	CUM.
1	VTR	OPENING	.30 ✓	
2	STUDIO (1) VTR + CAPS	Link into HIGHLAND GATHERING	3.50	4.20
3	STUDIO (1) (2 & 3) + CAPS	Link into INTERVIEW WITH HAMISH McGUINESS	4.30	8.50
4	STUDIO (1) VTR + CAPS	Closing link into 15" VISION OF HIGHLANDS 55"	1.10	10.00

Putting it all together. In a programme like this, there is no point in backtiming. Sequence 4, comprising the closing link and VTR, has fixed times, which means that we must come out of the interview at 8.50.

TIME CHART : THE AFTERNOON PROGRAMME

SEQUENCE	EST.	CUM.	REAL	ON AIR
1. OPENING	0.30		30	30
2. Link into HIGHLAND GATHERING	3.50	4.20	4.10	4.40
3. Link into INTERVIEW	4.30	8.50	4.10	8.50
4. Closing link into VISION OF HIGHLANDS	1.10	10.00		

Closing box

@ 9.00 ROLL VTR / 1' to end programme 1'
@ 9.05 GO GRAMS / S/B END CREDITS
@ 9.30 30" to end programme 30"
@ 9.45 15" to end programme 15"
@ 9.50 Counting out of programme in 10 . . 9 . .
@ 10.00 Off air

Step-by-step guide
We shall start at one minute to transmission.

PA does	*PA says*	*Script*
Starts digital countdown clock having already liaised with presentation re. on air time and overall duration	'One minute to transmission, one minute'	THE AFTERNOON PROGRAMME
Possibly the director rehearses the opening of the insert items in which case she would count them down		
	30" to transmission, 30" Standby VTR-2 with opening title	S/B VTR
	(VTR-2 beeps in response)	
Looks up to check ident on monitor displaying VTR-2	Thank you VTR	
Picks up overall stopwatch	15" to transmission, 15"	
Counts down from ident clock on monitor. Picks up insert watch	10" . . . 9 . . . 8 . . . 7 . . . 6 . . . ROLL VTR-2 . . . 4 . . . 3 . . .	ROLL VTR-2 – 5"
(Transmission screen goes blank)	2 . . . 1 . . . zero	
Starts both watches		VTR-2 OPENING TITLES
Puts master overall watch down Watches txn monitor and insert watch	On VTR for 30" Standby studio	DUR: 30" S.O.VT
Counts off insert watch	15" left on VTR. Camera 1 next.	S/B CAM. 1
	Counting out of VTR 10 . . . 9 . . . 8 . . . 7 . . . 6 . . . 5 . . . 4 . . . 3 . . . 2 . . . 1 . . . zero	

PA does	PA says	Script
Flicks insert watch back to zero and stops it while glancing at txn monitor to ensure VM has cut to studio. Notes down real time on r/o (Director cues Mary)		
		CAM 1 /MARY
Looks up to check ident on monitor displaying VTR-6	Stand by VTR-6 (VTR-6 beeps) Thank you VTR	MARY IN VISION — Good afternoon. In today's shortened version of the Afternoon programme, we will be bringing you a report from the gathering of the McGuiness clan, followed by an exclusive interview with Hamish McGuiness, self-styled King of the Highlands.
Counts down from ident clock. Picks up insert watch	ROLL VTR-6 4 . . . 3 . . . 2 . . . 1 . . . zero	ROLL VTR-6 — But, to begin with, John Tew reports from what must be one of the strangest gatherings in the western world./
Starts insert watch as picture appears	On VTR for 3'50" Standby first capgen 'Margaret McGuinness'	VTR HIGHLAND GATHERING
(At that pont the PA is handed a note saying that this insert runs for 4'10 She does not see any necessity to inform the studio because it would cause unnecessary confusion.)		DUR: 4'10" CAPGENS: Margaret McGuiness @ 1'05 Colin McGuiness @ 2'25 out words: "in all of Scotland"
She watches txn monitor as she is now not sure whether capgen timing is correct . . . it is!	Coming to capgen in 5 . . . 4 . . . 3 . . . 2 . . . 1 . . .	

(@ 1'10" into item)
3 minutes left on VTR, 3'

She tells the producer at the end of this item the programme will be overrunning by 20". The producer says to cut the interview down by 20"

(@ 2'10" into item)
2 minutes left on VTR, 2' standby for capgen

She gives a quick check that the name of the capgen is correct

Coming to capgen in 5 . . . 4 . . . 3 . . . 2 . . . 1 . . .

PA tells director decision to cut interview by 20" then presses switch talk-back key and tells presenter . . .

Mary, you've got just over four minutes for the interview

(The presenter gives a thumbs-up sign to show she has understood.)

(@ 3'10" into item)
1 minute left on VTR, 1 minute. Out words are: 'in all of Scotland'
Standby studio/standby camera 1

(@ 3'40" into item)
30" to studio, 30"

15" to studio, 15"

PA does	*PA says*	*Script*
Watches monitor as well as watch	Counting out of VTR 10 . . . 9 . . . 8 . . . 7 . . . 6 . . .5 . . . 4 . . . 3 . . . 2 . . . 1 . . . zero	
	On camera 1, 2 next	CAM 1 /MARY MARY IN VISION — With us in the studio we have Hamish McGuiness, known to his clan as King of the Highlands./
Notes down real time on r/o and works out final box	On 2. As directed interview next	CAM 2 2-s — Hamish, where does the title come from?
(The director cues in the capgen of Hamish as soon as there is a single shot of him)		AS DIRECTED INTERVIEW ON CAMS. 2 & 3 Capgen: Hamish McGuiness Duration: 4'10 approx.
At this point the PA knows that she must come out of the interview at 8'50" in order to get off the air on time. She should use her master overall watch to give the countdowns		
	(@ 5'50" into the programme) 3' left on the interview. 3'	
	(@ 6'50" into the programme) 2' left on the interview. 2'	
	(@ 7'50" into the programme) 1' left on the interview. 1'	
	Standby VTR/Standby grams	S/B VTR / S/B GRAMS
	(VTR responds)	

Action	PA	Script
She checks that the ident clock is correct	Thank you	
	(@ 8'20" into the programme) 30" left on the interview. 30"	
	(@ 8'35" into the programme) 15" left on interview	
	10" wind up on interview	8'50 end interview
Glances down at overall watch and notes time. They are running to time, thank goodness!	5 . . . 4 . . . 3 . . . 2 . . . 1	CAM 1 /MARY
	VTR next	MARY IN VISION — Well thank you very much indeed. And sadly, that's all for today. We'll be back at the same time tomorrow with a rather different view of our national
Counts down into VTR. Perhaps starts insert watch at zero but most probably would stay now with master watch	ROLL VTR . . . 1' to end programme. 1' . . . 4 . . . . . . 3 . . . . . . 2 . . . . . . 1 . . . GO GRAMS / S/B END CREDITS	ROLL VTR — sport. Meanwhile we'll leave you with some music and pictures of the misty Kingdom of the Highlands./ 9'05
	On VTR for 55" Standby with closing credits	VTR VISIONS OF THE HIGHLANDS — CLOSING CREDITS
(The director cues in the closing credits)	30" to end programme. 30"	DURATION 55" — Presenter: Mary Calloway Director: Len Greene Producer: Philip Smith
	15" to end. 15"	
	Counting out of programme 10 . . . 9 . . . 8 . . . 7 . . . 6 . . . 5 . . . 4 . . . 3 . . . 2 . . . 1 . . . zero. We're off air	10'00

13 On location – outside broadcasts

To separate outside broadcasts from other areas of television is, in some respects, misleading. The work of a PA on a live outside broadcast is very much defined by the nature of the programme. If the OB is of a variety show then her job would be very similar to that done by a PA working on the same type of show based in the studio. The same is true of many other outside broadcasts. Timing, rolling in pre-recorded items, counting in and out, logging, shot calling . . . all these elements might be and often are, involved. But there are instances where the work of the PA on a live outside broadcast differs from that of a PA working on a live studio-based programme and this is especially true in the areas which are very much the province of the live OB: sport and the large one-off event.

So what are the specific tasks to be undertaken by the PA in the OB scanner?

Living in the future

In a live studio, the PA keeps everyone informed as to where they are in the running order of the programme. She counts through each item and warns the studio what is coming next.

On a scripted programme the PA will also talk the studio through the show largely by means of shot calling from the director's pre-planned camera script.

On a live outside broadcast which has no script and is largely dictated by the event itself, the PA's prime contribution is to provide a running commentary of the event, not by describing what is happening at that moment but by talking through what is to come. This is known as 'living in the future' and its importance to the production is as follows:

Importance to directors

Many directors vision mix for themselves when working on a live OB. But whether they do their own vision mixing or not, their entire attention will naturally be taken up with what is immediate, what is happening at that moment. Their concentration is directed towards selecting the instantaneous shot and the one that is to come next. They rely on their PA for all the usual duties carried out by a PA in the gallery in addition to acting as a second pair of eyes by previewing the monitors which they are unable to concentrate on at that moment, especially when it is a large OB with many outputs.

Importance to camera operators

Isolated on scaffolding towers, positioned precariously on window-ledges or on rooftops, camera operators rely on the PA as their lifeline. They need to have warning of what is to come, of whether they are required to find a close up of a person or an exhibit and where that person or exhibit is to be found. They also need to know when the event has moved away from them giving them the opportunity to relax, especially in an event lasting some hours.

Importance to commentators

The man or woman in the commentary booth providing live commentary of the event is in a peculiarly vulnerable position. Many of them will welcome the advance information given by the PA.

Let us imagine that the commentator is about to talk about a rare floribunda at a flower show. The PA ought to be able to direct the camera operators to the correct plant. This would involve knowing what the plant looks like; knowing the layout of the show; it would mean having plans in front of her so that she knows which section of which stand in which part of the grounds the species is to be found. Then she will need to know which camera is closest to the plant.

But isn't that the director's job, you might well ask. Yes, but at that precise moment the director is engrossed in the immediate shots covering the judging of the roses. The commentator's talk will be the next item.

Timing

Timing on a live OB can be critical, not so much in terms of getting the programme off the air, although that can be a priority, but often the timing within the OB itself can require a fine appreciation of the event and an ability to 'feel' the situation.

On sports events timing might or might not be important depending upon the sport and depending upon whether the OB is being transmitted as an entire programme or being inserted into a studio programme containing other live elements.

Timing within the sports event, however, might be essential. For example, how far into the match did the goal occur, or how long to go before the end of the boxing round? In wrestling matches the time-keeper might well be listening for the PA's count and she would therefore need an insert watch for each individual round and the interval as well as a master overall watch.

Physical stamina

Along with the ability to squeeze herself into a tiny space, cramped confined and surrounded by people and equipment – if you are large or suffer from claustrophobia then steer clear of OBs! – the PA needs a certain amount of physical stamina especially when working on a live OB over several hours. She needs to be dextrous too in that there are far more telephones, keys and switches than in the studio gallery.

Communications

Communications form a vital part of the PA's job. She will be liaising constantly with everyone associated with the OB. She should never assume that information has been passed from one person to another but should always check for herself.

This again is standard PA work but whereas in the studio there is an advanced communications system with a full-time back-up staff of engineers, on OBs there will be a system – often ingenious in its complexity but sometimes tenuous in that it has to rely on cables stretching across muddy fields, Post Office land lines that can be subject to interference or breakdown, cameras perched in remote spots and electricity supplies dependent upon the whims of generators. And instead of a back-up staff of unlimited numbers, all she will have is a small band of dedicated, harassed engineers who have been up since the small hours trying to get it all to work!

Leaving nothing to chance

A PA's job on OBs therefore relies very much on all the standard skills of the PA plus a close involvement in the programme content and an extra large dose of being able to adapt to anything at a moment's notice.

Because of the uncertain nature of most live OBs, few PAs working in them will leave anything to chance if they can check it beforehand. On a complicated variety show held one New Year's Eve, show business personalities, singers, dancers, politicians, anyone in the public eye and in London that night was to be brought into the studio or interviewed in the street during the course of the evening. It was to be one of those relaxed, unscripted programmes that look so simple when watching at home and are so complex if you are involved in the transmission.

For the PA, however, there was one detail that was of great importance and that was the necessity of cutting to a shot of Big Ben in time to see the great clock begin to chime prior to midnight. Because the cut to that shot was

'. . . ready to time Big Ben.'

critical, on the night before transmission she was to be seen on Westminster Bridge at ten minutes to midnight, stopwatch in one hand, notebook in the other, ready to time Big Ben.

For any PA likely to need the same information, she found that at −20″ to midnight there are run-up chimes which last for precisely 10″. At −10″ there is silence lasting for 10″ and the first chimes are, as one would expect, on the stroke of midnight precisely.

14 On location – single-camera shooting

Continuity

Having mentioned the need for continuity when shooting single camera, film-style technique, just what is it?

What is continuity?

Continuity means being aware, both before and during the making of a programme of the final edited version, no matter in what order the material is recorded. It means trying to ensure, as far as possible, that when the programme is edited in its final form it will flow from shot to shot and scene to scene in a smooth manner with no continuity errors to distract the viewer.

Importance of continuity

The importance of continuity is greater than the popular misconception that is only to do with hats and levels of drink in glasses. Continuity is not something that is solely confined to drama. A simple interview, a documentary, a magazine programme . . . whatever the content, so long as the technique of recording is by the use of one camera, shooting out of sequence, then continuity, to a greater or lesser extent, is required.

Doing the job

To go into the job of continuity in any great depth is outside the scope of this book and has been fully covered in my book *Continuity in Film and Video,* (Focal Press, 1989). But in this chapter I will try to set down the essentials of the job.

Observation

The PA must watch the action of each shot and make relevant notes about what seems to be important in terms of matching shots. The information she builds up should be used during the recording to ensure that continuity of action, dialogue and props will be preserved from shot to shot and scene to scene.

What is important within a shot?

That, for anyone trying to do continuity on whatever type of programme, is the all-important question. First of all it must be realized that it is just not possible to watch *everything* within a shot. Neither is it necessary to do so. The art of continuity lies in the specialized and not generalized observation that is undertaken and there are a number of fairly simple guidelines that can be followed.

Framing

Without knowing the size of shot you cannot begin to do continuity. In a very wide shot, for example, it is not necessary to be as observant of all the actions as it would be in a close shot. It is generally true to say that the more that is happening in a shot, the less continuity matters. But, conversely, in a shot that does not have a lot of action, perhaps a close shot of two people, the smallest movement will be important. In shooting on videotape you have the advantage of being able to look at a monitor which will give the precise framing of each shot.

Largest moving object

Once the shot size is clear then the largest moving object within the shot is the most important to observe as a viewer's attention will be drawn to it – whether it is a head seen in close up, a person seen in mid shot or a large elephant entering frame and dominating the wide shot. Another general rule relates to colour. In the same way that the audience will tend to notice the largest moving object in the frame, they will also notice the brightest colours.

The main characters

If there is a great deal happening in the shot then you should concentrate on the main characters. Stick to watching the people the programme is about, the people who, theoretically, should hold the audience's interest and attention. Other people in the shot only assume importance *in relation to* the main subjects.

Given the above guidelines, what else does one need to notice?

Screen direction

Screen direction comes under the umbrella of continuity. Exits and entrances should always be noted and the information used to match shots preceding, following and cutting in to the master shot.

'The largest moving object within the shot is the most important to observe as a viewer's attention will be drawn to it . . .'

It is easy to become confused, especially if a number of different angles relating to the same scene are being shot. It helps to draw a quick, rough sketch of each shot, noting the camera position. When referring to exits and entrances always do so from the camera's point of view, which is also that of the audience. For example: 'Tom enters frame camera left and exits frame foreground right'. The convention is that if an object or person is to appear to be travelling in the same direction during a number of consecutive shots then they must always cross frame in the same direction.

A car is driving along the road. The camera is positioned so that the car appears to be travelling from left to right. Any subsequent shot intended to cut directly on to the first one must show that car travelling from left to right of frame. It is also permissible to show the car travelling straight to camera or directly away from camera. What will not work, however, is a shot of the car travelling from right to left. Cut directly on to the first shot, the car would appear to have changed direction.

Crossing the line

The principle of screen direction applies not just to movement. John and Jane are sitting opposite each other. John is on the left of frame, Jane on the right. These positions have been established in the master shot. Any other shot cutting directly on to or in to that master must feature John on the left and Jane on the right otherwise it would appear that the two people have changed places. The 'line' is an imaginary one drawn between people's noses as they look at each other. Providing the camera does not cross that line, shots that cut directly on to each other will match.

The only exceptions to this convention are as follows:

- When the camera moves during the course of a shot to a fresh position establishing Jane on the left of frame and John on the right.
- When the people move. Either John or Jane gets up and moves around to a different position.

Action in relation to dialogue

Another basic aspect of continuity is one which is often overlooked. It is, nevertheless, one of the most important points because it directly affects the editing process. Continuity of action, and especially action in relation to dialogue, is vital. Actions should be repeated at the same point on each shot and if dialogue is involved then it assumes even greater importance.

In working on scripted dialogue you should use the script as the basis of your notes and peg down the actions at the relevant places.

Costume

You should note the costumes worn by the actors in each shot and check that the same clothes are worn where there is direct continuity between shots and scenes.

Props

Props can be divided into two distinct groups, dressing and action.

- Dressing props, i.e. furniture etc. used to dress the set, do not, by and large, get moved (unless they are shifted in order to re-position camera and/or lights) and it is only necessary for you to have a general plot of these props.

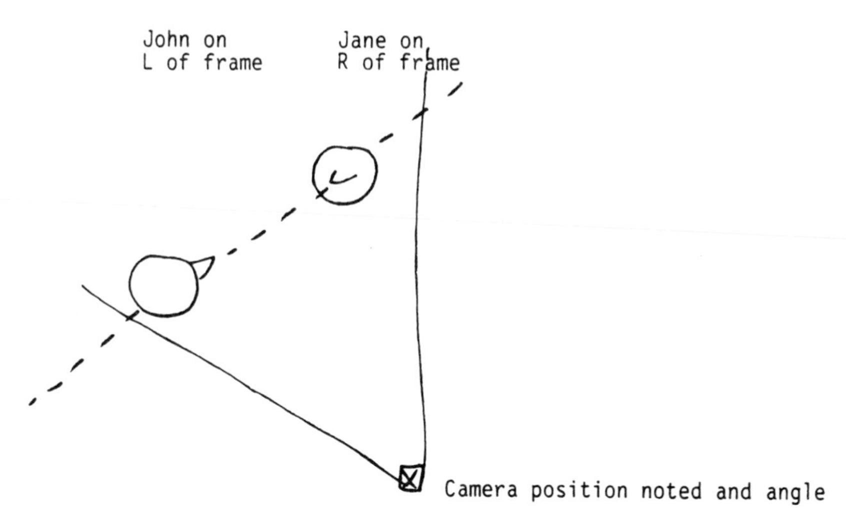

Dot in a line between people's noses as they look at each other. Providing the camera does not cross the line, shots that cut directly on to each other will match: i.e. John and Jane will remain on the same side of frame as in the original and will not 'jump' in frame.

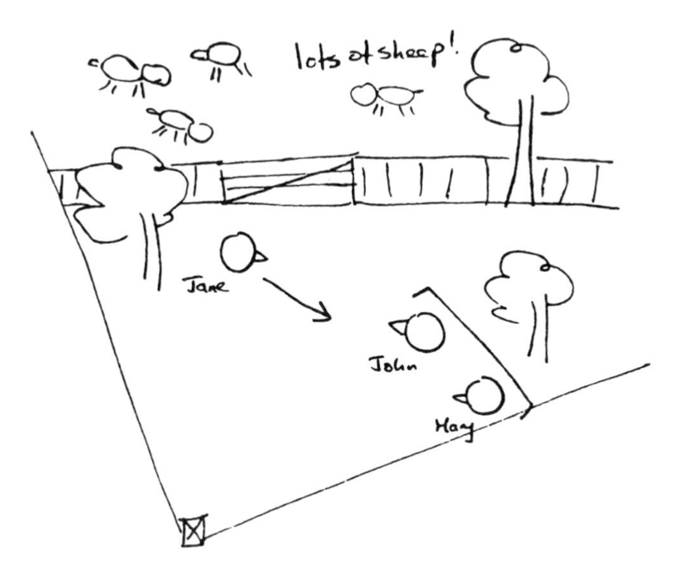

Always do a rough diagram of the shot, showing the camera position as well as the actors and main props. It does not matter how rough your diagram is, it will be of enormous help in matching future shots.

- Action props are those used by the actors during the course of the action of a scene. It is therefore much more important that you know what these props are, where they are placed, when they are picked up or moved and at what point in relation to the dialogue.

Continuity is not in any sense a precise discipline. You might have noticed all the relevant things within a shot, your notes might be models of neatness and your drawings exquisite *but* there are other factors involved:

Artists

With the best will in the world – and most actors will try to do it right – some artists are not good on continuity and even the best will have 'off' days. Actors are people – not robots and that is something you should never forget.

Directors

Directors too, are people, and even if you have noted wrong continuity it is by no means a foregone conclusion that the director will retake for that purpose. There are many reasons – many of them entirely valid – why a director will accept a shot which you feel to be wrong by virtue of continuity. There is nothing you can do about it.

Drawings

Always try to reduce the scene happening before your eyes to a simple sketch or drawing. It is far easier to relate to a sketch than to half a page of written description. The sketches need not be works of arts, indeed they are better if they are quick, rough and to the point, but they will be invaluable.

Keep close to the camera

Finally, in watching what is going on, keep as close to the camera as possible so that you are seeing the action from the same viewpoint. In addition, hold your script up on a level with the action in order to minimize the time between watching and writing. And if you can have a monitor somewhere within sight in order that you can check the framing then so much the better. Don't forget that while it may be possible to spool back and re-run shots for continuity, it is an immensely time-wasting operation.

Part Four: Post-Production

1 Videotape editing

In the 1950s, videotape was edited by cutting and splicing the master recorded tape. These edits were very 'hit and miss' as there was no way of previewing an edit decision before cutting up the tape. Sound editing was a further problem as the sound was about 9″ in front of the picture. The videotape that was used in broadcasting organizations was two inches wide and known as the Quadruplex system.

The development of electronic editing in the early 1960s meant that the master tape was no longer physically chopped up. Instead the material was re-recorded on to a second tape, reassembled into its final programme order. This system is known as 'dub' editing and is the basis of all linear videotape editing done today. Early electronic editing was a fairly haphazard system, despite the introduction of Editek in 1963 which controlled the edit point electronically.

When time code was introduced in 1967 enabling a more precise control of the edit point than ever before, videotape editing became more efficient and accurate.

The establishment of one inch videotape with a helical scan arrangement in 1978 meant that it became possible to stop the tape and inch through the action. There is a picture of sorts as you spool through at high speed. With modern one inch machines, broadcast quality pictures are available at any speed from −100% (i.e. normal speed backwards) through a freeze frame to +300% (three times normal speed forwards).

The later developments of computer controlled systems has meant that the director and videotape editor can enjoy even more sophisticated, precise control of editing than is available to the film editor.

Film editing

In film editing using negative/positive film stock, the shot material is processed and is in a negative form. From this negative a positive 'rush' print is made. The sound recorded on location is transferred to magnetic film of the same gauge as the picture (16 or 35 millimetre), and, after synchronization by the film editor's assistant, the 'rush' print and synchronized sound are viewed and then broken down into individual shots.

The editing is done by joining these shots together in whatever order is decided upon by the director and editor. Shots can be experimented with in any order, they can be lengthened, shortened, replaced or re-cut at any point. It does not matter what state the print gets into: the print is only a guide for the negative cutter who splices the negative *only* after all the creative decisions have been finalized. This allows immense flexibility. It means that changes can be made at any time until the negative is cut. Once the negative is cut the transmission print or prints can be made without loss of picture quality since they are made directly from the picture negative.

Videotape editing

Until the advent of digital VTRs, the restriction upon the number of generations one could go to without loss of picture quality meant that ideally the editor should get it right the first time, not necessarily in terms of shots but in terms of the precise overall length of the programme.

It would be possible to leave a gap of, say 20″ in the middle of an edited programme knowing that you had a shot of exactly 20″ to drop in, but if you wanted to drop in a shot of only 15″ then either all the edits from that point to the end of the programme would have to be re-made – a tedious and time-consuming operation – or you would have to go to another generation of tape, i.e. copy all the shots up to the 15″ shot, edit in that shot and then copy the edited end section.

In an entirely digital editing environment, successive tape generations hardly matter but the cost of an all digital suite is so expensive that with complex programmes it makes much more sense to perform an 'off-line' edit before the final 'on-line' edit.

Off-line editing

In its crudest form, off-line editing might entail making a VHS copy of the original material with 'burned in' timecode and, using a two-machine VHS editing suite, roughly join together the basic elements of the programme, i.e. some interviews, in such a way as to ensure that when the programme is actually edited, the narrative will flow as intended and the running time is approximately correct. The PA might be required to log the various timecodes resulting from this off-line edit in readiness for the on-line edit.

In its most sophisticated form, off-line editing could mean that the videotape editor finalized a complex drama production to the satisfaction of the director, producer, and anyone else who was involved before proceeding to an on-line edit that was not attended by any of these people.

In essence, the advantage of off-line editing is that you are not tying up expensive broadcast quality equipment while making the creative editing decisions and you are also not constantly running backwards and forwards over the precious master tapes. Providing the equipment is capable of providing reasonable quality pictures over successive generations, any linear tape format can be used from VHS through S-VHS and U-Matic to Betacam SP (see Part Four chapter two for an explanation of these formats).

'These edits were very "hit and miss".'

The editor will use one of a number of logging systems such as Cuedos, Shotlister or Turbotrace to produce an EDL (edit decision list) which he can take to the on-line edit on a computer disc. In an ideal world, the editor who performs the off-line edit should, of course, carry out the on-line edit since he or she has made the creative decisions on the material. If this is not possible then the off-line editor would normally be present at the on-line.

In place of a linear tape format a non-linear editing system such as Avid or Lightworks may be used for the off-line edit. These systems hold the edited and unedited material on magneto-optical computer discs rather than conventional video cassettes.

On-line editing

Many programmes do not need any off-line editing and sometimes it is not feasible or possible for off-line editing to take place. In that case the videotape editor will edit the master tape direct from the recording with no stage in between.

If the PA attends this editing, she should take her notes from the recording and, if a scripted programme, her coverage script marked up as explained in Part Three chapter four.

Video conforming

If all editing decisions have been completely finalized in an off-line editing situation and no creative decisions are to be made at the on-line editing stage, then the on-line editing is often known as video conforming, since each shot is edited solely by reference to its time code and not to its picture content.

Booking videotape editing facilities

If you are responsible for booking editing facilities then you should, wherever possible, discuss with the videotape editor precisely what is needed. If, for example, mixes are needed then a suite with a minimum of three machines will have to be booked unless you are using a digital format offering pre-read. If the director requires special effects, you need to ensure that the edit suite provides a DVE which offers the sort of facilities likely to be needed. You may also need special provisions for stereo.

It is not necessary that you should have the technical expertise to know exactly what is required for your programme – it is sufficient for you to know that there are many areas in television which are both complex and intensely specialized and for which you need expert help.

Sound

Sound is so often the poor relation to picture in television. It is wrong to think of it in this way; sound can bring pictures alive by the use of music and different effects can add immeasurably to the overall creative effect.

Sound is normally recorded simultaneously on to videotape along with the picture, alteratively it may be recorded separately and interlocked by means of timecode. For example, some scenes in a single camera drama might be recorded on a DAT machine to give greater flexibility. The master sound from a rock concert is likely to be recorded multi-track and mixed down later. In either of these cases guide sound might be available on the original videotape.

Depending upon the complexity of the sound content of the programme, the sound may be mixed by the videotape editor at the on-line edit if he or she has the facilities available. Modern digital tape formats can carry up to four soundtracks but the advent of stereo sound in television effectively means that these are reduced to two. If the sound post-production is complex a separate dub may take place after the editing.

What is a dub?

A certain amount of confusion arises over the use of the term 'dub' which can have three separate meanings:

- making a copy of an existing audio or video recording,
- re-voicing dialogue, e.g. 'I saw a French film last night which had been dubbed into English',
- 'dubbing', 'dubbing session' or 'post-production dub' are all names given to the process of mixing together the various sound tracks – location sound, commentary, music, library sound effects and any post-synchronized material – to form a master sound track for the programme.

If there is to be a sound dub at a later stage in videotape post-production, the actuality sound will be laid off from the master videotapes at the time of the on-line edit. Before the dub can take place, a copy of the picture is made which carries the same timecode as the master edited videotape.

In the dubbing theatre (which might be called an audio post-production house or even an audio sweetening suite!), additional sound will be laid, using the copy of the master videotape as a guide. When all the sound likely to be required has been assembled in this way, the various audio tracks are mixed together to form the final sound track. Timecode of the edited programme is displayed which relates to notes made on the dubbing cue sheets produced by the assistant videotape editor or audio assistant.

The equipment used for handling this sound will vary from conventional timecode synchronized analogue tape recorders (for example, the Nagra-T quarter-inch machine, or Studer Multi-track), to magneto-optical disc based systems such as Audiofile and Augan. DAT machines are also used and the final mixed soundtrack is very likely to be in the form of a timecoded DAT. Some audio post production houses undertake to lay back the final sound track on to the master videotape, alternatively it is taken back to the videotape editor for lay back.

As the variety of audio post-production systems increases, it is essential for the PA to ensure that the on-line editor is aware of the way in which the sound dub is to be conducted. It is probably advisable for him to talk to the sound supervisor directly.

Stereo

A word of warning about stereo. By implication stereo is two-channel sound. However, the two tracks involved may either be recorded M and S or A and B.

Location sound is more likely to be M and S – the M standing for mono and the S for stereo. In this case, a special microphone is employed which gives a high quality mono signal and a separate stereo or width signal, the level of which can be controlled in post production.

A and B sound means, in domestic terms, left and right. Most, but not all, final mix sound tracks will be A and B.

If the videotape editor is required to produce the final sound mix at the on-line edit, he will need equipment to decode M and S location sound tracks to A and B.

When making shot lists, it is essential that the PA notes whether stereo sound is recorded M and S or A and B.

2 The PA's guide to videotape

(All you ever wanted to know about videotape but were afraid to ask . . .)

Television systems throughout the world

Television systems throughout the world are not standard. There are basically three colour systems. There is the 625 line PAL system, used, for example in the UK and Europe (excepting France), Australia and China; the 525 line NTSC system, used, for example in the USA and Japan, and the 625 line SECAM system, used, for example in France and the CIS. All these systems, at present, have an aspect ratio of 4 × 3, i.e. the shape of the television screen is in the proportion of three units up to four across. Future television developments such as PAL-plus and high definition television (HDTV) are likely to have an aspect ratio of 16 × 9.

Material recorded on one television system as opposed to videotape format is not compatible with another system even if the make and type of machine is the same, unless the material is played through a standards converter.

With cable and satellite and the general opening out of television in the UK has come a world-wide marketing and exchange of programmes. While this would not concern the PA working in any of the large television companies, it is important for PAs in smaller ones and especially those working in satellite and cable to appreciate these differences.

Types of videotape

There are in existence quite a bewildering array of videotape formats and sizes. New products are arriving on the market so quickly that the comprehensive list that follows may be incomplete by the time you read it.

Analogue and digital

The simplest way to understand the difference between analogue and digital is to think of a domestic audio cassette tape and a compact disc. The audio cassette is an analogue recording and the compact disc is a digital one. The CD gives spectacularly better quality. Digital video recording gives a similar improvement when compared to analogue video recording.

Most of the new formats developed in the last few years have been digital rather than analogue. Apart from higher quality, the main advantage of digital VTRs is that there is minimal loss of quality when pictures are copied, as in editing, compared to analogue VTRs. In addition, many digital VTRs offer a facility known as read-before-write, which means you can replay an existing picture off a VTR at the same time as replacing it with a fresh one. This is especially useful in editing applications involving complex multi-layering, but is best explained by the example of adding name superimpositions to an already edited programme using one machine only. The image is replayed (read) off the machine, combined with a character generator in a vision mixer and re-recorded (written) on the same place on the tape.

Although to get the maximum advantage from a digital VTR it would have to be used in an entirely digital environment, i.e. digital camera, digital vision mixer in the edit suite and so on, there is still a spectacular improvement when digital VTRs are used in an otherwise analogue environment.

A further complication arises as it is possible to work component or composite in either analogue or digital.

Component and composite

A composite signal, such as from a 1″ VTR, utilizes one wire to carry all the picture information. A component signal, such as Betacam SP, uses three wires, one for the black and white information and two for the colour information.

While component is a better way to process television signals, giving better quality pictures, it is not a convenient way to distribute those pictures to the viewer. All the national terrestrial transmissions are composite systems – PAL in the UK and most of Europe, SECAM in France and the CIS, and NTSC in America and Japan.

Videotape formats

All the following are *linear,* i.e. they involve magnetic tape moving continuously past a rotating head drum. *Non-linear* video recordings are an entirely new development and, at present, are mostly found in off-line editing systems such as AVID and Lightworks. Future developments may give broadcast quality pictures from non-linear sources.

Analogue

2″ Quadruplex videotape (composite)

This was invented by Ampex in 1956 and was the standard tape used for broadcasting throughout the world for over twenty years. Nowadays 2″ VT would only be encountered by the PA when working with archive material. Most large broadcasting organizations retain a machine for replaying archive tapes and there are specialist facility houses who undertake transfers of Quad tape to modern formats.

A 2″ machine will not give a recognizable picture until it has run up to speed which takes ten seconds, and one cannot see any sort of picture when spooling fast. If 2″ material is to be incorporated into an on-line edit, it must first be transferred to a modern format since today's editing systems cannot control 2″ VTRs.

Many programmes recorded on 2″ will not have any timecode and if the PA is required to log shots from a 2″ programme without timecode, she should ask the tape operator to set the machine's tape counter to zero at an agreed point on the tape and make a note of that point. She can then use the machine's counter to log the material that needs to be transferred. When the material is transferred to a modern format, timecode should be requested for the new recording and this timecode should be noted by the PA in her editing notes.

1″ Videotape (composite)

1″ videotape superseded 2″ throughout the broadcast industry but is now itself being superseded by digital formats. 1″ videotape allows for still frame and slow speed replay together with a picture of sorts when spooling. Most 1″ machines have timecode.

2″ and 1″ videotapes are wound on open spools but all other videotape is contained in cassettes.

¾″ High band U-Matic (composite)

High Band U-matic spearheaded the change from film to tape for news gathering. Marginally regarded as broadcast quality it has now largely been superseded for broadcast purposes. It is, however, much used in lower budget corporate productions and as an off-line editing format. SP U-Matic is an improved version of High Band and, like High Band, may or may not have timecode.

¾″ Low band U-Matic (composite)

Never regarded as broadcast quality in the UK, the cassettes are identical to High Band cassettes. High Band and Low Band systems are not interchangeable although cassettes of either format can be played back in the machines of the other format for viewing purposes only. Either way round, the picture is in black and white. Some U-matic machines, however, are switchable between High Band and Low Band.

Low Band is likely to be found in off-line editing systems and, being a robust format, in continuously repeating playback systems found in exhibitions etc. Timecode is not standard on Low Band but may have been incorporated in individual systems by, for example, sacrificing an audio track on the tape.

½″ Betacam and Betacam SP (component)

Betacam SP is a vastly-improved development of the original Sony Betacam format, and has virtually become the world-wide medium for single camera acquisition. Depending which system, NTSC, PAL, etc. is in use, cassettes holding up to thirty-six minutes of material can be contained in a recorder mounted directly on the back of the camera itself. While Betacam cassettes can be replayed in Betacam SP machinery, Betacam SP cassettes can only be replayed in the older Betacam VTRs if these have been specially modified.

Betacam SP is a very high quality analogue component format but is not particularly robust and can suffer from drop-out. Timecode is standard on all Betacam and Betacam SP machines.

½″ M2 (component)

Introduced as a rival format to Betacam, it never found quite the same popularity as Betacam SP and although some major television companies in the UK adopted it as a replacement for 1″, it is now more likely to be encountered in professional, rather than broadcast, environments. Timecode is standard on M2 machines.

½" Super VHS

S-VHS achieves a compromise between composite (using one wire) and component (three wires) by using two wires; one to carry the luminence (black and white) part of the picture, and the other to carry the chrominance (colour) part of the picture. It is used by some organizations for news-gathering and will be found in low-budget corporate productions. It is also a high-end domestic format. Some machines have timecode.

Hi-video 8 (8mm wide)

A technically similar system to S-VHS, using different width tape. Hi-8 would be used in similar situations to S-VHS. Some machines have timecode.

Video-8 (8mm wide) (composite)

Mainly found as a domestic camcorder format.

½" Betamax (composite)

Almost obsolete domestic video format.

½" VHS (composite)

Domestic video format used world-wide. Also used extensively in television for viewing purposes and low quality off-line editing.

Digital

At the time of writing, the following digital formats were either available or projected. By the time you read this book other formats may have appeared and the projected ones may or may not have materialized. This is, perhaps, the fastest changing technical area of television. All digital VTRs have timecode as standard and at least four audio tracks.

¾" D1 (digital component)

Developed by Sony, this is the highest quality VTR format in current use. It is used in applications such as title sequences where very many tape generations are required since the high quality makes these generations imperceptible. It is also used for television commercials post-produced on videotape, following a high quality digital telecine transfer from 35mm negative film. D1 VTRs are extremely expensive which is why they are not generally used in broadcast work.

¾" DCT (digital component)

Standing for Digital Component Technology, DCT is claimed by Ampex to be the industry's first practical digital component production system available from one manufacturer.

¾″ D2 (digital composite)

Marketed by Sony and Ampex, D2 looked set to be the standard replacement for 1″ tape throughout the industry. The arrival of D3 has, however, tended to divide the market.

½″ D3 (digital composite)

Developed by Panasonic, this format offers similar quality and features as D2 but on a narrower width tape. A D3 camcorder is available. This format has been adopted by some companies, including the BBC, as a replacement for 1″.

½″ D5 (digital component)

A projected format from Panasonic, D5 will be a component version of D3. D5 machines will be able to replay D3 tapes. In the UK, Channel 4 has announced that it will adopt D5 as its transmission format when it moves into its new premises.

½″ Digital Betacam (component)

By the time you read this, Sony will have introduced a digital version of Betacam which will also be capable of replaying analogue Betacam SP tapes. Sony claim that the price of digital Betacam VTRs will be closer to that of Betacam SP machines than that of D1 VTRs.

The distribution chain

In the days of 2″ videotape, what you did after you finished the programme was simple. In the UK, you broadcast it from an analogue composite (PAL) transmitter mounted on a tower fixed to the top of a hill. Nowadays this sort of broadcast (no longer 2″ videotape!) is termed 'terrestrial' to differentiate it from satellite or cable.

Cable broadcasting, which is not extensive in the UK, but is widely found in the USA, involves direct connection to the viewer's home and is based on the national television system – in the UK, PAL, in the USA, NTSC etc.

Satellite television was first available in the UK from two sources: SKY, a PAL system, and BSB, which used a component-orientated system known as D-Mac. Financial circumstances caused the amalgamation of BSB with Sky to form BSkyB which standardized on the PAL system.

Future broadcasting will involve high definition television, wide screen television and a host of other improvements to transmission. In the UK we already have digital stereo sound (NICAM) on terrestrial transmissions. We shall therefore see an even greater variety of technical standards and systems. All this will have immense bearing on the format of videotape used in programme-making.

3 Information for the editor

Single camera shooting

Single camera shooting requires some form of record to be kept so that the editor and director have accurate information about what has been recorded when they come to reassemble the material in its correct order.

Unscripted programmes: shot lists

On an unscripted programme, for example a documentary, this shot list would be the only written information about the recording that exists, so it is extremely important that it is written both carefully and accurately.

If the budget does not allow for the PA to be on location then she might be required to compile a shot list from a VHS copy of the recorded material with burned-in timecode. This is very unsatisfactory, through no fault of the PA, and is an example of cost-cutting which frequently ends in additional expense as vast amounts of time can be wasted in editing in order to allow the director to hunt through the tapes for the correct shots.

If the PA has not been on location, for example, all she can note down is 'W/A mountain' against the timecodes of differing shots of mountains. It might, however, be extremely important to know which is the shot of Mount Ararat and which Mount Sinai. If the director was unable to make extensive notes on location then any subsequent shot lists must be inaccurate, especially if compiled by someone who was not on location at the time.

The shot list would contain the following information:

1. PROGRAMME TITLE, NUMBER, DATE OF RECORDING
2. TAPE ROLL NUMBER
3. LOCATION DETAILS
   You should note down the location, whether interior or exterior, day or night. If exterior you should note the weather.
4. VISUAL IDENTIFICATION OF THE SHOT
   If there is time code, the 'in' point at the start of each shot should be noted down. On a highly disciplined programme such as a drama shot film style, a clapper board may be used. The advantages being:
   - it gives those people who are more used to working on film the same sense of security,
   - there is a sense of discipline in using a clapper board which is never acquired by just running the camera. The whole unit is made aware that a shot is about to be taken.

- on sophisticated programmes where the master sound is not recorded on the same tape as the video, it gives instant confirmation in post production that it is in sync, even when the actual synchronization is achieved electronically.

5. SHOT DESCRIPTION
   An accurate description of the shot is the next essential. A list of the most commonly used shot descriptions and their explanations is given in Part Two chapter eight. A basic shot description would suffice for a scripted programme but if unscripted a fuller description would be helpful, i.e. who was in shot, how they entered frame, how they exited and so on.
6. INTERVIEWS
   When compiling a shot list of an interview it is important to note down the questions asked by the interviewer and give an outline of the reply. It is also helpful to note the time-code at the start of each question.
7. NAMES
   Ensure that you write down, with the correct spelling, any names and titles that may be required for capgens.
8. TAKES
   Some of the most important details to note relate to the number of takes for each shot. If the action of the shot is unsatisfactory for one reason or another, it is repeated again and again until the director is satisfied.
   A note should always be made of the number of takes relating to each particular shot, the good takes marked and the reason given why the others were considered unsatisfactory (NG).
9. TIMING
   You should take a stopwatch timing of each take of each shot. Time the action only, from when the director says 'cue' to when he says 'cut' or 'stop recording'. This timing will be useful in the editing and also for building up a cumulative running time of the programme.
10. SOUND DETAILS
    Note the sound roll number if not recorded on the same tape as the video picture. If more than one sound track is recorded, give details of the split, i.e. radio mic. Track 1, FX mic Track 2, mono, stereo A and B, stereo M and S etc. (see page 204 for explanation of these terms).
11. WILDTRACKS
    It is particularly important to make a list of wildtracks and what sequence they go with. A wildtrack is a sound recording without picture. It could be a voice-over or a recording of sound effects or atmosphere. For example, a sequence is shot in an art gallery consisting of fifteen or more shots of pictures and details of pictures. If the edited sequence uses the actual sound taken as each picture is shot the sound will change on every picture cut, which is undesirable. A good sound recordist will not record sound during the actual shooting – thus enabling the director to talk to the cameraman – but will record a continuous wildtrack of sufficient duration to cover the whole of the edited sequence. If, as is likely, sound is recorded

onto the VTR, the camera will have to be running. The sensible thing to do in this case is to ask the cameraman to take a rough shot of the microphone during the wildtrack which enables the editor to find it easily when spooling through the tape.

Scripted programmes

On scripted programmes, drama for example, you would provide basically the same information presented in a different way as you have the advantage of being able to provide the editor with a script.

1. RECORD OF SHOOTING
   This record, which is similar to an abbreviated shot list, could be typed out as one long list or it could be presented in the film fashion of putting each shot on a separate sheet of paper or small index card.
2. COVERAGE SCRIPT
   You should provide the editor with a marked up coverage script, each line denoting a shot and the shot identified by means of the 'in' point of time code and the videotape roll number.

All this information is primarily for use *after* recording has been completed. Any notes, drawings or photographs taken *during* the recording which relate to the day-to-day continuity of matching one shot with another while recording is in progress are of no interest to the editor.

Multi-camera shooting: recording log

The PA should compile an accurate log of the recording by noting down the following:

1. TAPE ROLL NUMBER
2. TIME CODE 'IN' POINT
3. SHOT NUMBERS
   Taken from the camera script of the recorded section.
4. EPISODE AND SCENE/SEQUENCE NUMBER
   Identifying what, in the script, is covered by this section of recording.
5. TAKE NUMBER
   This should correspond with the verbal ident given by the floor manager.
6. NOTES
   Anything that was wrong with the recorded section should be noted down and any additional notes or reminders for the editor should be written in this column.
7. RUNNING TIME
   The running time, taken from the stopwatch, should be written down.

SHOT LIST

"STATELY HOMES" Dir: Peter McDonnell 21st February 1986

Programme number: 21Z/367942/B VTR No: 31548

Location: Withyton House, Staffs

T/C 'IN'	TAKE	SHOT DESCRIPTION	DUR.	OK/NG
10.05.00	1	EXTERIOR: HOUSE & GROUNDS Wide establishing shot of house from gates - looking down tree lined drive	.45	OK
10.12.05	1	Z/O from house to WS a/b	1.00	NG - person in shot
10.16.20	2	a/b	1.05	OK
10.32.45	1	Start wide and z/i to house	.20	NG - cam
10.43.40	2	a/b	.55	OK
		EXTERIOR : SUNKEN GARDEN		
11.55.10	1	WS panning L-R to gazebo	.40	NG - cam wobble
11.57.28	2	a/b	.42	OK
12.14.48	1	WS panning R-L (starting on gazebo)	.47	OK
		INTERIOR : DRAWING ROOM Interview with Lord and Lady Corley		
14.35.20	1	2-s Lord and Lady Corley on sofa Q: Lord and Lady Corley, this beautiful house is an amalgamation of many different styles of architecture. When was the first building constructed? A: (Lord C) There has been a house on this site since Elizabethan times...west wing.. fire...restoration in Jacobean times..bits added by each succeeding generation. Q: And have you added anything? A: (Lady C) The vegetable garden and sunken garden altered	6.40	OK

Example of a shot list.

Coverage script

It is helpful to the editor and director to provide a marked-up coverage script, (see example on p. 140). This should show a line for each take.

The sound dub

If there is to be a sound dub, the PA might be required to find library effects, tapes or discs for the dub. During the final run through of the master sound track the PA should time any music for copyright purposes.

Specially composed music/commentary

If music is to be composed specially for the programme, or if there is narration or commentary to be written, a copy of the master edited tape will have to be organized, probably on VHS and sent to the composer/writer. This copy should have time code burned in. When the time comes to record the music or commentary, if this is to be done in a separate sound session from the dub, it is important to arrange for a copy of the master edited videotape. The PA should check which VT format the sound studio can handle. This copy should not only have the time code *displayed* in the picture area for the convenience of cueing, but also carry an actual *recording* of the time code so that music or commentary can be recorded on to a tape which is synchronized with the picture as in the dubbing process.

Commentary script

If the PA is required to type a script for the commentary it should be typed in treble spacing, with the time code 'in' points (of the edited tape time) on the left hand side. A wide margin should be left for notes and alterations. Sentences should never be carried forward from one page to another. When typing the time code, it is often more convenient for the hours and frames to be left off.

TIMECODE	COMMENTARY
00.02	The sad story of the East and West Junction and South Wales Mineral Railway is an object lesson in how the entrepreneurial spirit of the Victorians did not always flourish.
01.59	Conceived in 1874 by a flamboyant Welshman, Alun Thomas, its ambitious scheme was to link the blast furnaces in South Wales with iron ore from Northamptonshire mines after local stocks were used up.
02.45	Sadly, the expected traffic never materialised and by 1901 the Company was in a sorry state.

Example of a commentary script.

4 Clearing up

Once the programme has been recorded and the editing and dubbing have taken place, it is by no means the end of the production for the PA. There now follows a positive deluge of paperwork to be done, before the PA can throw the left-over scripts into the wastepaper basket (after ensuring that some are retained for posterity) and banish the programme file to whatever vault, basement or building houses dead files.

Even on live programmes the PA has a certain amount of clearing up, although that is usually minimal compared to the clearing up necessary on recorded programmes. What, however, is basic to all television programmes, whether live or recorded is a form itemizing in detail the entire content of the programme. This form is vital both as a written record for possible future use of the programme (a second transmission, overseas sales, cassette release for home sales etc.), for any enquiries relating to any part of the programme and for all details of copyright.

This form goes under different names, but the essential information contained therein will be the same:

Programme as completed (transmitted/televised/recorded)

General details

The name of the programme, the production or job number, the VTR number and roll numbers of other tapes must be given. The date of recording and/or transmission should be shown. Information on the studio or OB site used for the recording, the names of the producer, director and PA also come under the general details.

Content and contributors

A synopsis of the programme should be written and everyone appearing in it must be noted. If scripted, the name of the author (and adapter), address and telephone number and details of the writer's agent. The names of all the contributors, whether artists, presenters, extras or members of the public must be listed. The kind of contract the contributors were employed under, details of rehearsal and performance days in the studio or on location. The names of all musicians, the contract issued to them, the instruments they play, the dates of rehearsal and recording and the type of session for which they were booked. The type of music (theme, incidental, etc.) that was being recorded must be given. If any facilities were used, i.e. a stately home, location caterers, a preserved steam railway, these must be included on the form.

Copyright details

Any book or publication used in the programme must be noted, together with the name of the author, the date of publication, the publisher and whether or not the book is in copyright. The same information is required for book illustrations. Any still photographs used must be written down: whether they are black and white or colour; their reference number and description and details of the copyright holder.

Film inserts

The title of the film used and its source, i.e. whether it was specially shot for the programme or came from elsewhere, possibly stock shots from the company's film library or bought in. The original source of the material *must* be stated. It is not enough to use a piece of film from another production and then state the source to be the name of that production. Where did the earlier production gain the film from – in other words who owns the copyright? It is essential to trace film back to its original source in order to ensure that payment is made to the correct person. In addition, the gauge of the film used, i.e. 16mm, 35mm, the running time and whether sound or silent must be noted.

Videotape inserts

The same information is required for videotape inserts as for film, i.e. the tape reel numbers, the format, the source of the material for copyright reasons and the running time. Remember that if you used shots from a videotape of a library programme which were originally taken on film, then your programme contains film inserts even though no-one has actually handled any film. It is essential to trace the copyright back to its original source.

Music

For copyright information the following details need to be noted: the composer, publisher and arranger of the music; the performer, the title of the piece and, if on a record, which side and which band. The duration of the music used in the programme must be noted and a full description of its use: was it specially composed for the programme, is it background music, incidental music, visual or non-visual?

Other clearing up work the PA might have to do could include:

PROGRAMME AS TELEVISED

SERIES TITLE & NO. IN SERIES		PROGRAMME TITLE	PROD NO	VTR NO
DATE OF VTR	DATE OF TXM	STUDIO OR OB SITE	TAPE NOs	
PRODUCER	ELIGIBLE/NOT ELIGIBLE FOR RESIDUALS (Delete as applicable)	DIRECTOR	ELIGIBLE/NOT ELIGIBLE FOR RESIDUALS (Delete as applicable)	
PRODUCTION ASSISTANT		PA's SIGNATURE	EXT	DATE

FILM TITLE (AND SOURCE)	GAUGE	RUNNING TIME		STOCK OR SPECIALLY SHC
		SOUND	SILENT	

PROGRAMME AS TELEVISED 2
(Continuation sheet)

PROGRAMME TITLE	PROD. NO.	PLACE OF REHEARSAL	
RUNNING TIME	PRODUCTION ASSISTANT & EXT. NO.		
NAME OF ARTISTE	REHEARSAL DATES	PERFORMANCE DAY(s)	COMMENTS

PROGRAMME AS TELEVISED 3 – MUSICIANS

PROG. TITLE					PROD. No.			VTR No.			
Names and Addresses* of Musicians and Instruments played	Date	No. in Group or band	Rehearsal prior to Basic Session		BASIC SESSION						COMMENTS Incl. Session type** travel, subsistence, title of music when applicable.
					REHEARSAL		VTR/TX/SOUND RECORDING		OVERTIME		
			From	To	From	To	From	To	From	To	

PROGRAMMES AS TELEVISED

PROGRAMME TITLE			PRODUCTION NO.		VTR NO.	
MUSIC						
PERFORMER	TITLE OF MUSICAL WORK	COMPOSER	PUBLISHER	RECORD NO. & LABEL	DESCRIPTION OF USE State whether visual or background/vocal or instrumental	DURATION

Example of programme as televised form.

Final estimate of costs

Some companies require a final costing to be compiled by the PA.

Artists' fees/supplementary payments

In a large company, you might need to send a note to the finance department to state that work has been satisfactorily completed in order that the final payments can be made, or you might need to make the payments yourself. Supplementary payments, i.e. overtime etc., must be taken into account.

In the case of children being employed, the most careful notes have to be kept by the PA of the hours of work, rest and tuition for each child and these details must be submitted to the authority concerned.

Transmission form (programme timing report/videotape timing and continuity)

These forms – given different names by different companies – all relate to the transmission of the programme and are essentially for presentation. Firstly the programme needs to be identified clearly by its name, production number, which should appear on the VTR start clock, and videotape number. The recording and transmission dates are required. Details of the opening shots (both sound and vision) and the closing shots should be shown. The programme's overall duration must be stated, the duration of the end sequence and part timings for commercial breaks. Any other information that would be helpful to the transmitting of the programme should be given on this form, i.e. a moving or dramatic end, a sustained period of silence, the absence of pictures, a presentation announcement to be made and so on. Sometimes a synopsis and cast list is required.

Post-production script

A post-production script is usually required for archive purposes. This should be a marked up, totally accurate script, if the programme was scripted. It should be accurate in that it should be *post*-editing and not *pre*-editing. Cuts might have taken place in the editing for example and these should be reflected in the script.

If the programme was unscripted a script must be made with a time code column, a column for pictures and a column for sound.

Billings

This is the information for the published programme guides. It should contain general details about the programme: the title, the author, the adapter, the episode number and title, the transmission day, date and time and the VTR number. The type of programme, a contact for any queries, the cast list, production credits, a brief synopsis and points of special interest also need to be given.

Publicity

Publicity material for the programme might need to be organized by the PA and press showings arranged.

'Send the programme file to its final resting place.'

Co-productions

If the programme is a co-production then a great deal of extra work is generated. At every stage of the programme the PA should be aware of the involvement of others, for example, in the clauses to be included in artists' contracts. Copies of the final programme might need to be made for the co-producer(s), music and effects tracks made and a good deal of liaison would take place, probably on a PA to PA level.

Overseas sales of the programme, cable and satellite sales, cassettes for home sale, will all require a certain amount of additional work by the PA.

'Thank you' letters

Don't forget to send 'thank you' letters to everyone who has helped in the making of the programme, especially those outside television. They are always appreciated.

And finally . . .

Throw away a mountain of unwanted paper, send the programme file to its final resting place, clear out your desk, unpin the wall charts, take home the potted plants and everything else that made the office home for the last few months . . . and prepare either to start again on Monday morning with a fresh director, a fresh programme or start job hunting again . . .

And remember that the ghost of this production will haunt you for many months to come . . .

Part Five: The Wide World of Television

1 The broadcast industry

The freelance PA

As I wrote in the preface to this edition, the freelance PA has become the norm rather than the exception in today's television industry and the likelihood is that a PA could expect to work for a variety of companies on short-term contracts, rather than stay with just one large company for a long period of time.

So how does anyone become a PA nowadays? The normal channel is still via work as a secretary/general dogsbody/runner with a production company. Then, if the opportunity occurs, it might be possible to train as a PA.

Training of PAs

At the time of writing there is a distinct need for some form of standardized training of PAs in the industry. The job of a PA, as I have tried to point out in this book, is far more than being solely a super secretary. There are specific skills to be learnt which can only be fully assimilated in a practical way through experience.

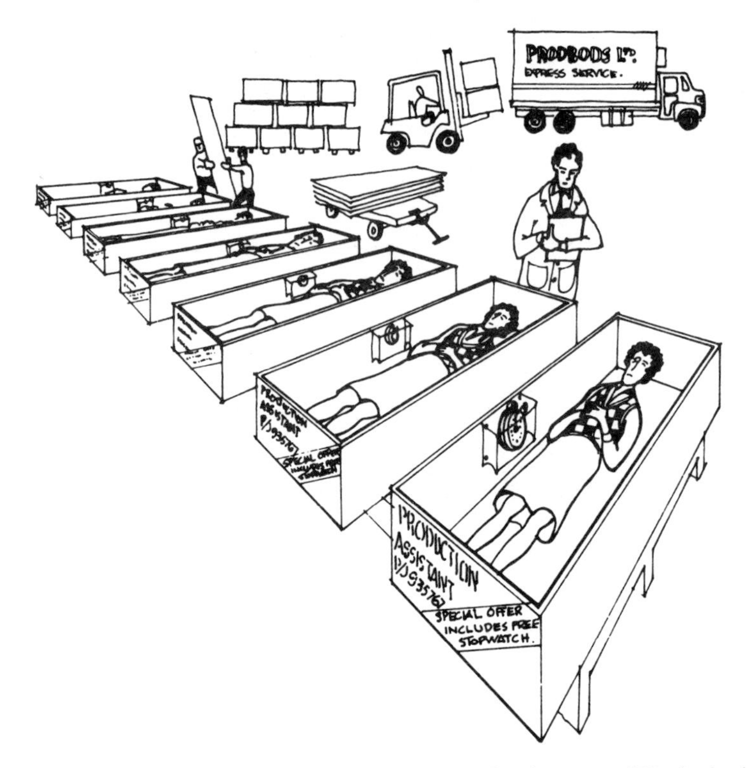

'There is a distinct need for some form of standardized training of PAs in the freelance sector of television.'

Finding a job

Having acquired training, how does one go about getting a job? Media jobs are advertised in certain daily newspapers and also in trade magazines for the industry. Then there are agencies which specialize in placing production staff. Failing that, you could try writing to companies directly.

Over recent years, a bewildering number of production companies and facility houses have sprung up – and more recently, a sad and increasing number of them have failed – and it might be helpful in terms of job possibilities, to try to group them as far as possible.

Facility houses

Facility houses do not originate programmes, they merely hire out specialized facilities, i.e. studios, editing, camera equipment etc. to individual groups, consortiums, or any body wishing to make a programme. The facility house will tend to specialize in just one technical area of the making of a programme. Facility houses do have their own staff: a group of videotape editors and

'A production company can be as small as one person.'

engineers might form their own facility house offering a range of post-production facilities, camera and sound operators might do likewise, but PAs are not usually in-built in this system. If a PA is required a freelance would be engaged.

Small production companies

A production company can be as small as one person – often a producer or director – with an idea for a programme. He or she might form a company, acquire a desk and telephone, perhaps engage secretary and get on with it. It can obviously be a great deal larger than that, but the small company would undoubtedly have to go to a facilities house in order to employ the expertise needed in making the programme. A freelance PA would be employed for the duration.

Large production companies

Some large production companies also operate as facility houses, providing facilities for others as well as making their own programmes.

Some of the companies employ permanent staff, including PAs: many of them operate solely with freelance staff, production people being taken on as and when required.

Companies that specialize

Some of the larger production companies tend to specialize in making one type of programme and all their resources go into the specific requirements needed to satisfy that format. This is following the trend in the USA where whole channels are devoted to one specific type of production. This means, of course, that the PA's work in that company will be wholly geared to that programme's specific needs. The difficulty for the individual PA arises when they feel limited within the company for whom they work but are unable to move elsewhere as they have never learnt the other skills needed by a PA.

Different types of television companies

There are different types of television companies.

Originator broadcasters

These are companies which, in the main, make and transmit their own material. They also buy in programmes made by independent companies and overseas companies.

Publisher/broadcasters

These are companies which transmit programmes but do not originate any themselves, except, perhaps, for local news. All other programmes are acquired from elsewhere.

2 The non-broadcast industry

Over the past few years video has expanded in all kinds of areas outside television. One has only to look at the video recorder in the home and the growing market for lightweight video cameras to record weddings, parties, holidays and baby on the lawn to see just how video has become integrated into people's lives – people to whom the processes that go into the making of a television programme would have been shrouded in mystery and mystique only a short time ago.

To meet this expanding market, cameras and equipment have become lighter, easier to handle and simpler to use.

We live in an age of instant pictures. Pictures are easier to assimilate than the written word and two generations have now been brought up with television as an indispensable item of furniture – an entertainer, a teacher, a friend. An explanation therefore of a new security system, of company policy, of different training methods, or simply how to work the new cooker that has arrived in the staff canteen is clearly better done by exploiting the potential of video. This thinking is not new. Some large industrial companies have had their own film units for years but the revolution in video technology has meant its expansion in the field of industrial training.

In other fields, too, the growth of video over the last decade has been astounding. In education, in schools and colleges, video has been used not just to make learning more pleasant and more relevant but the making of the programmes themselves has become an educational study.

In health, research and education, video has been found to be of inestimable value. Video is used in the training of doctors and nurses and training in any job or profession has been enhanced by the use of videotape to show real situations. Again, this is not new. Film has been used in this way for many years. But what is different about videotape is, firstly its relative cheapness in comparison with film and secondly instant playback. Students can see their own performances recorded on tape straightaway.

The latest developments in the use of videotape in training have been in interactive video, a process which incorporates computer assisted learning and computer assisted training with videos made on tape.

But videotape is not confined to training. Local governments record the changing face of architecture and environment on tape; the findings of archaeologists are recorded; conservationists use videotape to provide an archive of specific areas of natural beauty under threat and in furtherance of their cause; and in sales promotion and marketing videotape has made enormous headway. How often, for example, does one go into a shop to find a group of customers watching a monitor displaying an advertisement? Video has entered virtually every stream of life.

Does the non-broadcast PA exist?

In the burgeoning market covering such a wide field, is an animal such as a PA a necessity to any video production? Well of course in many ares where video is used, there is no need of a PA. There might be no need for a director either.

The simpler the use to which video is put the less need there is for a PA. In some non-broadcast areas the PA exists in the form of a secretary, general dogsbody and 'go-fer'. Sometimes the PA forms part of the company for whom the video unit is only one small section. The PA might exist under other names: production secretary, organizer, secretary and so on. Sometimes the PA is a freelance, employed specifically for a complex production. The needs vary. But in two areas at least there is a need for a PA and these areas we shall examine next.

'Video has entered virtually every stream of life.'

Corporate videos

Much of industry would go to a production company for its video needs. The company would be given a brief, and the client would work closely on the scripting and provide the locations and facilities necessary for the shooting. The production company would employ a freelance PA as required.

But some industrial concerns whose video output is so great have their own 'in-house' units. Some of these units are quite large and employ their own full-time PAs, but more usually freelance PAs are employed for a specific job.

What sort of work would the PA be expected to do?

The system

Every large company has its own style, its own way of doing things. This is true in television as well as other industries. Therefore the PA working in a video unit within a large industrial company will have to find out the system for getting things done and conform to it. It should not be too difficult as each company generally has written instructions pertaining to its system.

Large companies tend to be self-contained worlds with complex administrative procedures and it is of the first importance that the PA learns what these are. For example, booking overseas accommodation might be done through a network of overseas offices rather than in an ad hoc way.

Corporate image

Another thing the PA must remember is that the company will have its corporate image and that factors such as presentation, how you deal with people, public relations, even the way you and the crew dress will have a certain importance. Although, having said that, you must also remember that industry's perception of film or video units will be somewhat stereotyped and they might expect you to behave or dress in a more unconventional way than you would normally – you are, after all, 'artistic people'!

Different pressures

The time-scale for making corporate videos is different from making programmes for television. You are not working to meet a transmission deadline but you might be working to other deadlines, so there are pressures nonetheless.

For example, if you have arranged an interview with the chairman of the company, being such a highly paid individual, his time is precious and you come very low down on the scale of his priorities. Therefore the pressure will be to cut down on the time you have to interview him.

The priority is not the making of corporate videos, however worthwhile they are. The priority is the industry itself, whether it is banking, retailing, car

'You are, after all, "artistic people!".'

manufacture or whatever. So there will always be the frustration of knowing that you are low down on *anyone's* list of priorities.

Unlimited budgets?

Because we are in general talking about multi-million pound companies, most people labour under the delusion of unlimited multi-million pound budgets and their eyes gleam at the thought of making videos under such conditions.

The truth is often far removed. Videos are made to a specific budget in industry as anywhere else and the PA is responsible there as everywhere for the day to day float.

How videos originate

If a department wants a video made, a series of meetings will be held to get a clear idea of precisely what is needed. A director is brought in at an early stage, usually freelance. Sometimes a PA is not engaged until quite late in the setting up, so the director will overlap the PA's work.

Words are more important than pictures therefore more time is given to getting the script right and the script approval process takes a long time.

Once, however, the script is finalized and the dates arranged for the shoot, the PA is brought in.

The PA's job

The PA will be responsible for setting up the shoot. She will book the crew (freelance most probably), arrange travel, book accommodation. She might have to book artists but most likely that would have been done by the director if, indeed, artists are to be involved. The PA will work out the call sheets and the schedule in conjunction with the director and will type and distribute them.

Research

Because there is usually only the director and the PA at work on the setting up, the PA might well find herself involved in research relating to the video content. Some of the research might involve highly confidential matters within the company and therefore discretion is absolutely essential. Sometimes the research might involve simply finding archive film, music and so on.

On location

If the shoot is on location it will undoubtedly be single camera. If there is a studio either single or multi-camera will be used.

On location the PA acts as the 'go-fer'. She will log the shots, perhaps do simple continuity and provide an accurate shot description. Because only small crews tend to be employed, the PA would also do any necessary make-up, get coffee, arrange meals, keep people happy and sort out problems. She will hold the float and pay for accommodation, meals and travel. She will keep a note of the hours worked by the crew and she might well have to acquire props and costume. She will, in fact, do everything and anything necessary at the time.

Studio work

If the company has a television studio or is hiring a studio from a facilities company, the PA will most probably log timecode for editing. She might well have to type and operate the teleprompt machine. As with location work, the job will require flexibility and adaptability.

'She will hold the float and pay for accommodation, meals and travel.'

Post-production

The PA would normally be involved in the editing. After the recording the PA would follow the same clearing-up procedures as stated in Part Four chapter four. Copyright clearance would form a large part of her work.

Finally, the PA would ensure that the correct number of copies of the finished video were made. The numbers would vary. For a simple recording of a conference perhaps only one copy would be required for the chairman and two for the archives. But for a training or marketing programme the numbers might run into hundreds or even thousands.

Education

The role of the PA

Despite video being used so extensively in education, it is hard to be specific about the role of the PA. The person doing the job might well have other responsibilities, her main occupation being that of secretary to the producer or head of the video unit.

In a small unit there might only be herself and a producer. She would therefore be at the heart of any production, totally involved in every aspect. Anything that needed doing would therefore be shared between the producer and PA.

Studio and location work is likely to be similar to mainstream television but perhaps with extra responsibility for props, costume etc.

Public relations

A vital part of her work would involve dealing with people: with academics, students, professional people, and members of the public. Her suitability for the job would tend to be more dependent upon that aspect of her work rather than on the specific skills required in broadcast television.

Types of production

There is no limit to the range of productions that can be made – other than cost and resources – but the types can roughly be divided into two sections: those providing a service for the educational institute and those which could be classed as commercial contracts.

Academic productions

The demand would originate from the tutors and the aim would be to make recordings which were aids to teaching. The academic would ask the unit to make a programme on whatever subject, the script would be the responsibility of the academic and a series of planning meetings with the producer and PA would be held to work out the details.

Many of the productions required would be simply to record lectures for posterity. Many tutorials are now recorded on videotape, using a single camera and with no editing.

Recording interviews might be asked for. These interviews would be with various 'experts', whose views could materially aid the teaching of whatever subject is involved. The interviews would be recorded in a studio, 'as directed' and with no editing. How the final recording is used afterwards in teaching is up to the tutors: the aim is not to achieve a polished interview within a predetermined duration but to preserve on tape everything that is said by the interviewee.

'These productions usually have a strong academic content.'

More ambitious programmes could be and are undertaken. These could range from a full-scale drama production to a publicity film or a set-up 'mock' trial for law students.

The unit might also provide a service that students could use, i.e. viewing facilities, a tape library, the provision of portable equipment, even a studio for use on a self-access basis. Where film and television courses are run by colleges and universities, the facilities are used to give students varied studio experience and teach them the basics of television and film production.

Commercial contracts

Some colleges undertake commercial contracts that are strictly non-broadcast, providing they do not conflict with the required academic production work. These productions usually have a strong academic content and are often of a training character, i.e. 'How to handle the media' courses for local government personnel, educational programmes of a medical nature, the training of social workers through set up situations and so on.

Glossary of equivalent US terms

British term	*American equivalent*
Big close up (BCU)	Tight shot (TS)
Black edge generator (B/E)	Drop shadow generator
Camera script	Blocking script
Caption cards	Art cards
Caption generator (capgen)	Usually referred to as 'Kyron' or some other brand name of caption generator; sometimes losely referred to as 'graphics'
Chat show	Talk show
Closing title sequence	Closing credits, credits
Crabbing	Trucking
DFS	Electronic still store (ESS)
DVE/Quantel	Computer graphics, referred to by the brand name of the computer graphic generator
Floor manager	Floor director
Gallery	Control room
Gram	Record
Ident	Identification (ID)
'In' picture	First video
'In' word	Incue
Inject	Cut-in, remote
Insert	Piece
Lace up	Cue up
Lettering	Superimposition, supers – e.g. Chyron
Link	Copy leading up to a clip ('intro') or out to the next piece ('outro', tag)
Location manager	Field producer
Opting	Feeding – i.e., 'we're waiting for a network feed . . .
Organizer	Administrator, administrative assistant
Outside broadcast (OB)	Remote
'Out' word	Outcue, tag
Over-running	Running long
Post Office line	Telephone (land) line

Presentation	Traffic department; master control; network
Presenter	Newsreader, anchor
Production associate	Associate producer
Production manager	Unit manager
Production number	Account number
Recce	Reconnaissance
Rota	Rotation
Run-up time	Pre-roll time
Running order	Format
Sequence	Segment
SOT	Sound on videotape
Source sheet	Cut sheet
Standby	Filler, filler piece
Straightforward	On-camera, O/C
Studio	Studio session
Sub editor	Copy editor
Sum	Time
Superimposition	Supers, lettering
Talkback	Headset
Telecine	Film
Teleprompt	Prompter – both machine and operator
Tracking	Dollying
Transmission monitor	Air monitor
Under-running	Running short
Venue	Place, location
Vet	Edit
Vision mixer	Switcher – both machine and operator
VTR	Videotape

Index